BEHAVIOR OF MARINE ANIMALS

Current Perspectives in Research

Volume 1: Invertebrates

BEHAVIOR OF MARINE ANIMALS

Current Perspectives in Research

Volume 1: Invertebrates
Volume 2: Vertebrates

BEHAVIOR OF MARINE ANIMALS

Current Perspectives in Research

Volume 1: Invertebrates

Edited by
Howard E. Winn
Professor of Oceanography and Zoology
Narragansett Marine Laboratory
University of Rhode Island
Kingston, Rhode Island

and
Bori L. Olla
Head, Behavior Department
U.S. Department of Commerce
National Oceanic and Atmospheric Administration
National Marine Fisheries Service
North Atlantic Coastal Fisheries Research Center
Laboratory for Environmental Relations of Fishes
Highlands, New Jersey

PLENUM PRESS • NEW YORK-LONDON • 1972

Library of Congress Catalog Card Number 79-167675

ISBN 0-306-37571-0

A Division of Plenum Publishing Corporation
227 West 17th Street, New York, N.Y. 10011

United Kingdom edition published by Plenum Press, London
A Division of Plenum Publishing Company, Ltd.
Davis House (4th Floor), 8 Scrubs Lane, Harlesden, London,
NW10 6SE, England

Printed in the United States of America

CONTRIBUTORS TO VOLUME 1

Melbourne R. Carriker Systematics-Ecology Program, Marine Biological Laboratory, Woods Hole, Massachusetts

Hugh Dingle Department of Zoology, University of Iowa, Iowa City, Iowa

Brian A. Hazlett Department of Zoology, University of Michigan, Ann Arbor, Michigan

William F. Herrnkind Department of Biological Science, Florida State University, Tallahassee, Florida

Kenneth W. Horch Department of Physiology, University of Utah, Salt Lake City, Utah

Michael Salmon Department of Zoology, University of Illinois, Champaign, Illinois

Dirk Van Zandt Systematics-Ecology Program, Marine Biological Laboratory, Woods Hole, Massachusetts

PREFACE

What have been brought together in these volumes are works representing a variety of modern quantitative studies on a select group of marine organisms. Some of the species studied here represent basic biological experimental subjects–in some cases, marine versions of the white rat and pigeon–that are being used for a wide range of studies. Other species studied were virtually unknown as experimental animals.

The authors have studied their animals in considerable depth, often in both the field and the laboratory. It is this cross reference between real life and the artificial but controlled conditions of the laboratory which gives us the necessary understanding, and ultimately the means, for improving our rapidly deteriorating environment, a must for man's survival, maintenance, and improvement of the quality of living standards.

A direct outgrowth of a AAAS symposium entitled "Recent Advances in the Behavior of Marine Organisms" held in December 1966, these volumes include a reasonable balance between review and original unpublished research.

Of the many persons who have made these volumes possible, we wish to especially thank Nancy Fish, Lois Winn, Mabel Trafford, and Deborah Brennan. The latter two accomplished most of the final editorial work. The personnel of Plenum Press were cooperative in all aspects of our relationship. Only the two editors are responsible for defects in the volumes. We believe the papers presented are significant and will be of importance to members of the scientific community.

Howard E. Winn
Bori L. Olla

INTRODUCTION

The study of animal behavior may be viewed in a simplified way as an examination of the potentialities of the nervous system with its corresponding effector system. The fundamental starting point, whether the environment be aquatic or terrestrial, is observing what an animal does. To the behaviorist, this means observing the external expressions of the nervous system and effector organs as manifested in how an animal reacts to a particular stimulus. It is important to bear in mind that the data resulting from the direct observation of behavioral events, especially those occurring in the aquatic milieu, often contain a high degree of inherent random variation. Therefore, it is essential that questions be formulated and defined carefully and the data be collected and managed in a way that permits valid conclusions.

Marine animals and questions about their behavior are the main concern of the small but representative number of works appearing in these volumes. Until recent years, the main deductions on behavior of marine animals have been the result of surface sightings and net and hook collections correlated with measurements of environmental parameters. The number of collections and measurements depended, as it does today, on the ingenuity and budget of the individual scientist as well as the state of instrument development. Observations on marine animals in their natural environment have been encumbered to some degree by the fact that the observer has been physically unable to venture beneath the sea for more than a brief time. Over the past 15–20 years, the refinement of the self-contained breathing apparatus has extended the time and versatility of underwater observation. Improvements in manned submersibles, underwater television, and an array of remote sensing devices, have also contributed greatly to the opportunities for *in situ* study. Because it is not possible to control the myriad environmental influences in the sea, it is important to measure simultaneously various physical and biological parameters both above and below the surface. Subsequent correlation of these parameters with behavioral events permits meaningful interpretations and conclusions on cause and effect relations.

Relating laboratory findings to what is known of the animal's behavior in the sea is often a necessary step in interpreting and evaluating results. With all of its problems, the laboratory, because of the potentiality for precise control of most environmental cues, permits judgments about behavioral influences with less equivocation than is possible in the field. The ideal situation is one in which observations take place in both the field and laboratory with a continuous cross formulation of questions and integration of results.

There are certain requirements basic to laboratory study. An aquatic medium must be provided which closely approximates or duplicates natural sea water. Obvious parameters such as temperature, salinity, and various chemical balances must be constantly monitored and controlled to avoid variations in behavior caused by changes in the water medium. Confinement, as is the case with almost any vertebrate, be it terrestrial or aquatic, will often modify behavior in an unknown manner. Thus it is necessary to compare field and laboratory findings to determine what are real and what are laboratory-induced events.

The aim of this work is to present a variety of approaches and methods of analysis of both field and laboratory studies at various phylogenetic levels. Our goal is to show some small segment of the work that is currently being done in the area of marine animal behavior. The reader should, upon perusing the various chapters, note the interrelationships between organisms even though they are at different phylogenetic levels, and note the behavioral adaptations that permit the various species to be in harmony with the very specialized environment of the sea.

Most of the studies contained in this volume were performed on animals that live on the fringes of the sea. It is true that the accessibility of these species to the researcher makes them prime candidates for study. But more important than their availability is the fact that these organisms living in the estuaries and inshore areas are residing in the richest marine habitat. In contrast, the open sea is a veritable desert. Unfortunately, these inshore areas and their resident flora and fauna are highly vulnerable to man-made changes and destruction in the name of progress. Defining the destruction in a legal sense has been difficult because often no studies are undertaken until after the greatest damage has been done. Studies which concentrate on establishing normal patterns of behavior are necessary before it is possible to predict the outcome of particular man-caused modifications. Herein lies a valuable tool and asset in our attempt to salvage the environment, a tool that has not been adequately used.

Howard E. Winn
Bori L. Olla

CONTENTS OF VOLUME 1

Chapter 1

Orientation in Shore-Living Arthropods, Especially the Sand Fiddler Crab

by William F. Herrnkind

I. Introduction 1
II. Review of Orientation in Shore-living Forms 3
A. Amphipods 3
B. Isopods 7
C. Lycosid Spiders 8
D. Insects 9
E. Decapods 10
III. Orientation by Fiddler Crabs, Genus *Uca* 12
A. Orientation Mechanisms and Adaptiveness 12
B. Factors Controlling Preferred Bearings 18
C. Orientation during Ontogeny in *Uca pugilator*, the Sand Fiddler Crab 25
IV. General Discussion 38
Appendix 45
References 56

Chapter 2

Acoustic Signalling and Detection by Semiterrestrial Crabs of the Family Ocypodidae

by Michael Salmon and Kenneth W. Horch

I. Introduction 60
II. Patterns of Activity and Display 61
III. Signal Characteristics 64
A. Ambulatory Leg Movements 64
B. Rapping 66
C. Body Thumping 66
D. Stridulation 66
E. Ghost Crab Sounds 67

IV. Detection of Tonal Stimuli 68
A. Perceptual Capacities 68
B. Sound Reception 69
V. Detection of Rapping Sounds 78
A. Characteristics of the Visual-Acoustic System 78
B. Methods 79
C. Results 85
VI. Discussion and Conclusions 92
Acknowledgments 95
References 95

Chapter 3

Ritualization in Marine Crustacea

by Brian A. Hazlett

I. Introduction 97
A. The Concept of Ritualization 97
B. Methods of Study 98
II. Communication Studies in Marine Crustacea 99
A. Visual Agonistic Stimuli 99
B. Agonistic–Sexual Behavior Patterns 100
C. Cleaning Symbiosis 100
D. Tactile Stimuli 101
E. Chemical Stimuli 102
III. Model Presentations and Sequence Analysis 102
IV. Ritualization in Hermit Crabs 103
A. Agonistic Movements in *Petrochirus* 105
B. Film Analysis 109
C. Multilevel "Typical Form" in *Pagurus longicarpus* 115
V. Discussion 118
Acknowledgments 122
References 122

Chapter 4

Aggressive Behavior in Stomatopods and the Use of Information Theory in the Analysis of Animal Communication

by Hugh Dingle

I. Introduction 126
II. Information Theory 127
A. The Measure of Information 127
B. Measures of Communication 130
C. Some Comments on Terminology 133

III. Information Theory and Behavior ... 135
A. Methods ... 135
B. Applications ... 136
C. Limitations ... 138
IV. Stomatopod Aggressive Behavior—An Example of the Use of Information Theory in Behavioral Analysis ... 140
A. Description of the Behavior ... 140
B. General Methods ... 141
C. Intraindividual Sequences ... 142
D. Interindividual Sequences ... 143
E. Interspecific Sequences ... 146
V. General Discussion ... 149
A. Quantitative Analysis of Behavior ... 149
B. Information Theory ... 150
Acknowledgments ... 154
References ... 154

Chapter 5

Predatory Behavior of a Shell-Boring Muricid Gastropod
by Melbourne R. Carriker and Dirk Van Zandt

I. Introduction ... 157
II. Experimental Animals ... 158
A. Predator ... 158
B. Prey ... 160
III. Detection and Approach to Prey ... 160
A. Distance Detection of Prey ... 160
B. Approach to Prey ... 161
C. Close-Range Detection and Mounting of Prey ... 163
IV. Penetration of Prey ... 176
A. Selection of Borehole Site ... 176
B. Penetration of Shell ... 178
C. Duration of Boring Periods ... 209
V. Relationship of Size of Borehole, ABO, and Radula ... 219
A. Borehole Characteristics ... 219
B. Borehole, ABO, Radula, Shell Height, and Other Parameters ... 221
C. Variation in Size of Boreholes ... 226
D. Maximum Depth of Boreholes ... 228
VI. Feeding ... 230
A. Penetration of Mantle Cavity of Prey ... 230
B. Biting and Swallowing ... 231
C. Rate of Feeding ... 233

D. Gaping of Oysters 234
VII. Discussion and Conclusions 235
Acknowledgments 241
References 242

Index xix

CONTENTS OF VOLUME 2

Chapter 6

Development and Uses of Facilities for Studying Tuna Behavior
by Eugene L. Nakamura

I. Introduction 245
II. Observations and Research at Sea 246
III. Observations and Research Ashore 258
IV. Conclusion 273
References 274

Chapter 7

Regulation of Feeding Behavior of the Bicolor Damselfish (*Eupomacentrus partitus* Poey) **by Environmental Factors**
by Robert A. Stevenson, Jr.

I. Introduction 278
II. Methods 281
III. Results 283
A. Distance from Rock 283
B. Feeding Rates 290
IV. Discussion 293
Acknowledgments 300
References 301

Chapter 8

Daily and Seasonal Rhythms of Activity in the Bluefish (*Pomatomus saltatrix*)
by Bori L. Olla and Anne L. Studholme

I. Introduction 303
II. Materials and Methods 306
III. Results 307
A. Daily Activity 307
B. Seasonal Activity 311
C. Shifted Light 315

D. Constant Light 318
E. Retinal Rhythm under Constant Darkness 322
IV. Discussion 324
References 325

Chapter 9

Behavior of Symbiotic Fishes and Sea Anemones

by Richard N. Mariscal

I. Introduction 327
II. The Anemones 329
A. General Description 329
B. Ecology 329
C. Behavior 333
III. The Fishes 335
A. General Description 335
B. Ecology 335
C. Behavior 336
IV. Discussion and Conclusions 355
References 357

Chapter 10

Acoustic Discrimination by the Toadfish with Comments on Signal Systems

by Howard E. Winn

I. Introduction 361
II. Materials and Methods 362
III. Results 367
A. Physical Parameters for Facilitation 367
B. Pen Tests 370
IV. Discussion 372
A. Signal Parameters 372
B. Signal Systems 375
Acknowledgments 383
References 384

Chapter 11

The Effect of Sound Playback on the Toadfish

by James F. Fish

I. Introduction 386
II. Methods 388
A. Experimental Design 388

B. Analysis ... 390
C. Description of Playbacks ... 390
III. Results ... 396
A. Calling-Rate Experiment ... 396
B. Boatwhistle-Pattern Experiment ... 398
C. Cycle-Time Experiment ... 405
D. Antiphony Experiment ... 408
E. Tone-Pattern Experiment ... 411
F. Response-Time Experiment ... 415
IV. Discussion ... 418
Acknowledgments ... 432
References ... 433

Chapter 12
Using Sound to Influence the Behavior of Free-Ranging Marine Animals
by Arthur A. Myrberg, Jr.
I. Introduction ... 435
II. Mammals ... 436
III. Fishes ... 438
IV. Invertebrates ... 462
Appendix ... 463
Acknowledgments ... 465
References ... 465

Chapter 13
Visual Acuity in Pinnipeds
by Ronald J. Schusterman
I. Introduction ... 469
II. Underwater ... 471
A. Methodology ... 471
B. Results ... 475
III. Aerial vs. Underwater ... 477
A. Methodology ... 477
B. Results ... 479
IV. Luminance ... 481
A. Introduction ... 481
B. Methodology ... 482
C. Results ... 485
V. Summary and Discussion ... 487
Acknowledgments ... 491
References ... 491

Index ... 493

Volume 1: Invertebrates

Chapter 1

ORIENTATION IN SHORE-LIVING ARTHROPODS, ESPECIALLY THE SAND FIDDLER CRAB

William F. Herrnkind

Department of Biological Science
Florida State University
Tallahassee, Florida

I. INTRODUCTION

The motile macrofauna of coastal and estuarine sand beaches is represented nearly exclusively by arthropods, particularly semiterrestiral marine crustaceans. The permanent residents include amphipods (Talitridae; Hurley, 1968), isopods (Tylidae; Edney, 1968), and decapod crustaceans (Ocypodidae, Grapsiade, Mictyridae, Coenobitidae; Bliss, 1968). More occasional residents are wolf spiders (Lycosidae; Papi and Tongiorgi, 1963), beetles (Carabidae, Staphylinidae, Tenebrionidae; Papi, 1955*a*; Pardi, 1956), and mole crickets (Gryllotalpidae; Pardi, 1956). The groups successfully populating this environment do so in spite of the deleterious effects of variable and extreme physical conditions (Pearse *et al.*, 1928; Moore *et al.*, 1968). These include physical stresses posed by abrasion and removal of habitable substrate, or deposition of sediment, during storms. Physiological stresses, including insolation, osmotic imbalance, drying, anoxia, drowning, and poisoning by high concentrations of hydrogen sulfide, occur more regularly as a result of heavy rain, periodic tidal inundation, and aerial exposure. Biological pressure is also considerable, since the beach inhabitants are subject to predation from the land by mammals (rodents and raccoons) and toads (*Bufo*), from the air by shorebirds, and from the water by fishes (sciaenids) and portunid crabs, especially *Callinectes* spp. (Herrnkind, 1968*a*,*b*, and unpublished observations). The environmental conditions are such that Pearse *et al.* (1928) commented, "A marine sandy beach seems like an inhospitable place for . . . animals to become established."

The mechanisms adapting shore-living crustaceans to this environment are, for the sake of discussion, variously classified as morphological, physiological, and behavioral.* Morphological adaptations include gill chambers constructed so as to maintain moisture for aerial respiration, resistant integument reducing water loss by desiccation, and well-developed locomotory appendages enabling efficient migration to and from suitable refuge and feeding areas. Physiologically, various of these forms show thermoregulatory adaptations, osmoregulation, and mechanisms to accomplish molting processes in conditions of water shortage (Bliss, 1968). Nevertheless, none of the shore-living species can withstand exposure at one point on the surface in the intertidal zone by day for periods exceeding a few hours. Consequently, behavioral mechanisms are indispensable adaptations in these groups.

The three major behavioral mechanisms often interacting to ensure individual survival are shelter seeking, rhythmicity in activity patterns, and directional orientation. Shelter seeking includes burying, burrow digging, and entering suitable interstices, all of which reduce the chance of drying, overheating, abrasion, physical displacement, and capture. This behavior is sometimes triggered by immediate stimulus conditions preceding physiological stress. e.g., high light levels or dry substrate, but also occurs cyclically with tidal, daily, lunar, seasonal, or combined periodicities (Brown, 1961; Enright, 1963; Barnwell, 1968). The periodicities represent the interaction of external environmental stimuli with internal timing mechanisms. Photoperiod and tidal cues act as Zeitgebers, setting the phase of the endogenous "clock" which can continue to operate relatively independently of external influences. The regular cycling of behavior in species of *Uca*, the fiddler crabs, involves a tidal rhythm (i.e., inactivity below ground in burrows while the beach is flooded) then emergence at low tide (mainly by day) for feeding and social behavior. The fiddlers thus avoid the consequences of aquatic predators and displacement or abrasion by waves and currents. Shore-living amphipods and isopods are mainly nocturnal in their feeding movement, remaining inactive in moist regions by day and thereby avoiding exposure during the most intense drying period. In addition to benefiting individual survival, behavioral cycles also time and synchronize courtship and reproduction and facilitate other life processes consequential to shore populations.

Directional orientation, the third component of behavioral adaptation, often works in conjunction with others. Oriented locomotion is absolutely required by shore animals to attain or return to areas affording refuge and

* For extensive discussions, see *American Zoologist* **8**: 307–392 (1968) and **9**: 271–426 (1969).

food at appropriate times in an activity cycle as well as for relocating hospitable regions, or the beach itself, if displaced landward or seaward (Herrnkind, 1968*b*). An animal disoriented in either time or space on an intertidal beach will likely perish.

All groups of shore-living arthropods facing the stresses mentioned demonstrate orienting mechanisms permitting attainment and maintenance of appropriate environmental conditions. In fact, nearly all groups mentioned share one mechanism—a sun compass enabling shoreward orientation under drying conditions and/or landward movement upon immersion (von Frisch, 1967, pp. 438–451). However, developmental processes, microecological requirements, and organismic characteristics of each group are greatly different, as shown by the following examples. Isopods and amphipods develop directly in brood pouches, while decapods have planktonic larvae. Amphipods and isopods have relatively simple behavior patterns and restricted movements, while wolf spiders and some decapods have complex behavior and extensive associated locomotory activities. Wolf spiders, ocypodid crabs, and insects demonstrate more advanced visual capabilities than amphipods and isopods (e.g., higher acuity and form recognition). Wolf spiders and insects are subject to drowning but resistant to drying, while the crustaceans are resistant to drowning but highly susceptible to desiccation. These various characteristic differences are reflected by differences in oriented behavioral activities, in the external and internal aspects of orientation mechanisms, and in the processes establishing orientation and directional preferences on a given shoreline.

The purposes of this paper are (1) to briefly review the knowledge of orientation by shore-living arthropods in the perspective of their ecology and life history, (2) to present new information on ontogenesis of orientation in a decapod, *Uca pugilator* (Ocypodidae), filling in an obvious gap in knowledge, (3) to compare and contrast orientational behavior in the groups discussed, particularly regarding the development and role of orientation under different microenvironmental situations.

II. REVIEW OF ORIENTATION IN SHORE-LIVING FORMS

A. Amphipods

1. Mechanisms and Adaptiveness

Talitrids generally occupy burrows by day in a reasonably well-defined moist zone near the high-tide mark (Williamson, 1951*b*), the exact location depending on the species, substrate moisture, and particle size (Bowers,

1964). They emerge at night and migrate either downshore as far as the midtide level (Williamson, 1951*b*) or inland up to 40 or 50 m to feed on detritus (Gepetti and Tongiorgi, 1967*a,b*). The animals reverse direction at or near sunrise, returning seaward or landward to zones of suitable substrate hardness and moisture content (Bowers, 1964). The seaward and landward bearings are more strikingly exhibited during the day, when amphipods are displaced from their burrow area. For example, individual *Talitrus saltator* landing on dry substrate orient seaward while those touching wet sand downshore move landward (Pardi and Papi, 1961). Their paths may be variable initially but usually straighten and remain so for distances of up to 80 m or until the appropriate zone is encountered (Williamson, 1951*b*). In overview, reliable directional orientation seems indispensable to talitrids during migrations and for return to habitable regions when displaced.

Several orienting mechanisms operate to guide landward and seaward directional escape responses in amphipods, although the most widely demonstrated and most studied is the celestial compass described in detail by Pardi, Papi, and coworkers (Pardi and Papi, 1952, 1961; Papi and Pardi, 1953, 1954, 1959, 1963; Pardi, 1954*a*, 1957, 1960; Papi, 1960; Ercolini, 1963*b*, 1964; Pardi and Ercolini, 1965; Pardi and Grassi, 1955). Compass orientation under the daytime sky is known in *Talitrus saltator*, *Talorchestia megalophthalma*, *T. longicornis*, *T. deshayesei*, *Orchestoidea corniculata*, *O. benedicti*, *Orchestia mediterranea*, and is probably present in some other talitrids. The presence of the celestial compass is demonstrated by the behavior of the amphipods in a level circular pan permitting test animals only a view of the sky. In this situation, the animals hop to the pan perimeter in the compass bearing corresponding either to landward on their home beach (perpendicular to the run of the shoreline where they are collected) if the pan is wet, or seaward if the pan is dry (Pardi and Papi, 1961), The mechanism is characterized as follows: (1) Orientation is generally perpendicular to the home shoreline; populations on different shores orient in accord with that particular configuration. (2) The orientation is a menotaxis, i.e., at any angle from 0 to 360°, relative to the guidance stimulus. (3) Equivalent orientation is performed on a basis of directional information from sun position or the plane of polarization from the overhead sky (by intraocular perception rather than by intensity differences), or both operate in conjuction. (4) The mechanism is time compensating, since the appropriate bearing is maintained during the day at any sun azimuth, and experimental phase shifting causes respective shifts in the bearing relative to the sky cues; e.g., a 6 hr advance or retardation of the light–dark period causes an approximate 90° change ($6 \times 15°$/hr rotation rate by the earth) in orientation angle. (5) The timing is independent of local external factors, since amphipods brought from Italy to South America adopted an angle to the sun appropriate to their original home coast.

Although this mechanism is well adapted to guiding escape responses by day, nocturnal migrations must depend on other cues or another mechanism(s). Nocturnal orientation by the moon is known in *Talitrus saltator* (Papi, 1960) and *Orchestoidea corniculata* (Enright, 1961). This is consequential both because the animals are nocturnally migratory and because the lunar orientation requires a different timing mechanism than the solar compass. The moon, of course, appears at different but predictable azimuths on succeeding nights, complicating navigation, although the rate of azimuthal change still approximates 15°/hr. Apparently, *T. saltator* endogenously compensates for the former over long periods while *O. corniculata* must reset its cycle each night. The basic visual mechanism, as in the sun compass, constitutes menotaxis to a distinct light source independent of moon phase (Papi, 1960; Papi and Pardi, 1963).

Orientation guided visually by landmarks occurs in the absence of celestial cues such as in fog, under overcast, and on moonless nights, according to Williamson (1951*b*). He reported that amphipods, such as *T. saltator*, apparently orient telotactically to landmarks, i.e., directly toward an object or area of optical contrast. The bearing taken, however, varies according to the shape of the landmark such that most paths lead toward apices of inverted triangles or to the foot of inclines. He suggest that this response produces landward orientation toward the silhouettes of sand dunes visible even at night. Some individuals of another species, *Orchestia gammarella*, orient consistently toward rocks projecting from the sand (at the beach level where the amphipods live) even in the presence of celestial cues. Landmarks, therefore, may be more influential to orientation than celestial cues under some conditions or in some species.

Nonvisual mechanisms are suggested in certain situations. For example, Williamson (1951*b*) found that individuals of *T. saltator* oriented landward even in dense fog, when the sky and landmarks are not visible to humans (and presumably not visible to amphipods). This of course does not exclude visual perception of ultraviolet cues, indicative of sun position, or infrared stimuli emanating from sand dunes inland or the sea. Further knowledge of responsiveness of crustaceans to natural levels of these stimuli would be helpful. Anemotaxis, orientation by the prevailing wind currents, also appears in *T. saltator* under appropriate conditions in the absence of visual cues (Papi and Pardi, 1953). Also, recent evidence by van den Bercken *et al.* (1967) suggests another nonvisual mechanism(s) for which no stimulus has been hypothesized. Here, individuals of *T. saltator* demonstrate weak but statistically significant orientational preferences for seaward and landward compass bearings under black rubberized cloth, apparently opaque to visible light, and on moonless nights. This is in contrast to the findings of Papi and Pardi (Pardi and Papi, 1953; Papi and Pardi, 1953) and Williamson (1951*b*), who report disorientation in similar situations. Moreover, the directional preference under nonvisual conditions is not statistically modified by eliminating the earth's magnetic field. This phenomenon requires further study, since the results thus far are equivocal, being based in part on statistical significance derived from large numbers of data points. More definitive orientation to celestial cues appears with far fewer data points.

2. Establishment of Preferred Bearings

The capacity of populations in different locations to orient escape responses perpendicular to their respective shorelines by the celestial compass is characteristic of the orientational behavior in the amphipods studied. This adaptive directional choice, according to available evidence, arises from so-called innate processes* without necessity of individual experience, according to Pardi (1960). Young amphipods (*Talitrus saltator*, *Talorchesti deshayesei*, and *Orchestia mediterranea*) born in the laboratory and raised under a fixed artificial light orient by celestial cues in the bearing appropriate to their parent's home beach. This presumably permits the young amphipods to orient properly immediately upon leaving the brood pouch. The direct development process, without free-living larvae, seems to preadapt amphipods for this capability (the effects of experience on young in the brood pouch are unknown).

The orientation demonstrated by young laboratory-reared specimens, however, is not wholly analogous to that of adults, since the cumulative directional preference varies noticeably as the sun angle changes, apparently as a result of antagonistic action by phototactic tendency. Furthermore, the dispersions of laboratory-reared amphipods 1–3 months old are greater than those of experienced individuals of the same age collected in the field.

* Because such terms as "innate," "instinct," and "learning" are subject to variable and controversial interpretation, I will use the terms "genetic," "maturational," and "experiential" in describing the nature of processes underlying the behaviors discussed. "Genetic" will refer to heritable characteristics passed from parent to offspring via the genome. "Maturational processes" are the developmental, physioanatomical changes occurring during ontogeny and growth. They are assumed to appear in all individuals under a wide variety of environmental conditions despite variable individual experiences. "Experiential factors" are those dependent upon individual experience in particular stimulus situations. "Learning" will refer to modifications of behavior as a direct consequence of experience. In cases where a behavior is first observed in a specific situation but the animal's previous responses in such situations are unknown, the behavior will be termed "spontaneous." These usages in no way imply that the processes are necessarily mutually exclusive during behavioral development.

This may suggest that experience plays a role in adjusting the somewhat diffuse innate orientation to actual environmental conditions (Pardi, 1960). Alternatively, environmental stresses may exclude any young unable to orient accurately early in life. Even so, the natural variation in genetically controlled directional preference would be advantageous in permitting succeeding generations to adjust to new shoreline configurations produced by long-term erosion and sediment deposition. Another possiblility is that experiential factors can fully dominate innate tendencies, but this raises questions, already apparent, of how specific directing mechanisms become incorporated into the genome. Certainly there should be further analysis of this unusual phenomenon and an attempt to determine the underlying mechanism.

B. Isopods

1. Mechanisms and Adaptiveness

Generally speaking, the tylid isopods on the beach have activity patterns and movements similar to those of amphipods. For example, *Tylos latreillii* along the Mediterranean become active nocturnally and migrate from burrows in moist substrate to inland areas up to 50 m away, then return to the lower zones before morning. Mass landward migrations are also induced by stormy seas or extended heavy rain (Tongiorgi, 1962, 1969). The Eastern Pacific species, *Tylos punctatus*, which lives in regions of more extensive tidal change, migrates nocturnally from the high tide line seaward, returning to the upper beach before dawn (Hamner *et al.*, 1968). In both cases, the major directional components of migrations are perpendicular to the run of the shoreline.

Escape movements are induced in these isopods, as in amphipods, by displacement to dry or wet substrate, whereupon they orient seaward or landward, respectively. The orienting mechanisms differ according to the species and relate to the ecological characteristics of the habitat. Individuals of *T. latreillii*, for example, orient by a celestial compass cued by the sun, polarized light, or the moon (Pardi, 1954*b*; Papi, 1960; Papi and Pardi, 1953). On the other hand, *T. punctatus* exhibits only a negative phototaxis to the sun and no apparent orientation to the moon, but shows a clear directional response in the presence of slopes as small as 3° (Hamner *et al.*, 1968). This species presumably can make use of gravitational cues, since individuals inhabit and traverse the lower beach where continuous wave and tidal action produce a consistent slope. Such orientation would be adaptively useless to *T. latreillii* because the beach surface above high tide, where they are most active, is often uneven, masking any general land—sea slope of the region.

The relative influence of genetic, maturational, and experiential factors on direction preferences in isopods is not known. One may infer that directional preferences in *T. latreillii* are inherited by reason of its direct larval development, as in amphipods. The slope response of *T. punctatus* is likely independent of individual learning since it requires no compass reference and is therefore adaptive to offspring on any beach.

C. Lycosid Spiders

1. Mechanisms and Adaptiveness

Several species of wolf spiders, particularly *Arctosa variana*, *A. perita*, and *A. cinerea*, living on sandy beaches and the banks of estuaries demonstrate a strong landward-directed escape response when they are blown, or fall, onto the water surface (Papi, 1959; Papi and Tongiorgi, 1963). Thus their orientational behavior is also adapted to the beach environment, although they are terrestrial forms. The primary mechanism of orientation is apprently a sun compass permitting maintenance of a constant compass bearing appropriate to their home shore (Papi, 1955*b*,*c*, 1959; Papi and Serretti, 1955; Papi and Syrjamaki, 1963; Papi and Tongiorgi, 1963; Tongiorgi, 1959).

The compass represents a time-compensating menotaxis to the sun and/or polarized sky light similar in operation to that in amphipods. Perception of guidance cues, particularly polarized light, is predominantly mediated by the primary eye pair (anterior median) but may normally involve additional visual information from the secondary eyes (anterior lateral, posterior median, and posterior lateral). Celestial orientation in the proper landward bearing can be effectively accomplished entirely by the principal eyes, which are sensitive to polarized light and are capable of mediating form vision (Magni *et al.*, 1962, 1964). However, orientation is more accurate if the fixed posterior median secondary eye pair, also sensitive to polarized light, is used as well. Celestial orientation cannot be performed by the secondary eyes alone.

Landmark orientation occurs in wolf spiders when celestial cues are obscured. Under such conditions, the spiders orient to optically contrasting regions (usually the shore). Adults apparently orient preferentially to celestial cues when landmarks are also present, although young specimens are strongly attracted to dark objects even under the clear sky (Papi and Tongiorgi, 1963). The young individuals, and perhaps adults as well, can orient landward nonvisually by following a line of maximum inclination (if one exists) on a sloping shore, as do some isopods.

2. Establishment of Preferred Bearings

The processes underlying the establishment of landward bearings oriented by celestial cues are primarily experiential in wolf spiders (Papi and Tongiorgi, 1963). Young *A. variana* hatched and reared in the laboratory initially demonstrate a generally negative phototaxis to the sun, or random orientation. After several days, the negative phototaxis subsides, and a weak northerly bearing appears spontaneously at all times of day, indicating the presence of a time-compensating menotaxis without reference to any particular shoreline; i.e., offspring show a northerly preference independent of the parent's home shoreline, Young spiders reared under the natural sky in simulated habitats consisting of a sloping bank of damp sand, with or without water at the bottom, readily adopt the celestially oriented landward bearing within 2–3 days. The factors likely affecting learning of the landward preference are locomotory movements up and down the steep slope, avoidance of water, or responses to humidity levels. However, the learning mechanism remains to be analyzed in detail. In summary, the northerly preference of young spiders apparently results from the maturation of the neurosensory system, while the direction is eventually modified according to experience in a given habitat.

After establishing a strong directional preference in nature, wolf spiders maintain the bearing even if held for weeks or months indoors, although the orientation becomes less precise (Papi and Tongiorgi, 1963). Nevertheless, they can learn new escape directions (also weaker than the original) when held in simulated habitats with specific shore bearings. This interaction of learning with physiographic features is further demonstrated by the variability of orientation in spiders from different habitats. For example, individuals living on islets in exposed stream beds or in sand dunes more than 100 m inland show erratic orientation or no apparent preference, while those living on straighter banks or closer to the sea show stronger, landward-directed preferences. Thus wolf spiders develop directional escape responses in accord with the habitat characteristics they find themselves exposed to but modify the orientation as conditions change or if they move. This labile mechanism seems significant to the spiders, since they move about over a wide range of habitats in their lifetime, whereas amphipods and isopods are restricted to a narrow band in the intertidal zone.

D. Insects

Landward–seaward celestial orientation is known in beach-dwelling insects including the tenebrionid beetle, *Phaleria provincialis* (Pardi, 1956), carabids, *Scarites terricola*, *Omophron limbatum*, *Dyschirus numidicus* (Papi, 1955*a*) and the staphylinid, *Paederus rubrothoracicus* (Ercolini and Badino, 1961). The orientation mechanism follows the general pattern widely found in insects (Jander, 1963, 1965) but performs the special function of maintaining the animals in a restricted habitable region of the beach. Presumably, specific directional preference can be learned as in ants or bees (Jander, 1957; von Frisch, 1967).

E. Decapods

1. Mechanisms and Adaptiveness

Thirteen families of decapods have members semiterrestrial or terrestrial in habit (Abele, 1970; Bliss, 1968), although the main sandy beach forms belong to the Mictyridae (*Mictyrus*), Grapsidae (*Sesarma, Goniopsis*), and especially the Ocypodiae (*Uca, Ocypode, Dotilla*). Grapsids such as *Pachygrapsus, Grapsus,* and *Hemigrapsus* are inhabitants of rocky shores, the former two being semiterrestrial and the latter a shallow littoral and tide pool form. Members of the terrestrial ocypodid genera *Cardisoma* and *Gecarcinus,* as well as the terrestrial anomurans *Coenobita* and *Birgus,* migrate from far inland to the shore zone to release larvae, often with lunar or seasonal periodicities (Gifford, 1962; Bliss, 1968). Orientation and other important aspects of these migrations remain to be analyzed.

Generally, the shore dwellers live either in burrows (*Ocypode, Uca, Dotilla, Mictyris,* etc.) or in humid crevices between mangrove roots, under driftwood and rocks (*Sesarma, Pachygrapsus, Hemigrapsus, Coenobita*), within or just above the high tide zone. Those semiterrestrial forms lowest on the beach, such as *Uca, Dotilla,* and *Mictyris,* remain below ground while the beach is submerged and become active diurnally and, to a lesser degree, nocturnally at low tide (Crane, 1958; Altevogt, 1955, 1957; von Hagen, 1962). Others further upshore, such as *Ocypode* and *Sesarma,* may be active at all tidal levels day or night (Hughes, 1966) but especially at crepuscular periods. The oriented movement consists of travelling, from the burrow or lair, downshore to scavenge or actively prey, then returning upshore either to a general burrow area or region of suitable substrate as in *Uca* (Altevogt, 1955) or to a specific burrow as do some individuals of *Ocypode* (Cowles, 1908; Hughes, 1966). These crabs, when threatened in the open, generally run either seaward into the water (*Grapsus, Pachygrapsus, Ocypode*) or landward to any available refuge (*Uca*). Here, as in the general migratory movements, the orientation axis is perpendicular to the run of the shoreline. However, this orientational capacity is exceeded by *Uca* and *Ocypode,* both of which are capable of returning to a specific burrow, under some circumstances, independent of its location relative to the shore axis (Cowles, 1908; Altevogt and von Hagen, 1964; Hughes, 1966).

The major orienting mechanisms in the forms thus far studied are visual, including both celestial compass and landmark guidance. Tests in circular chambers permitting only sky cues show a time-compensating celestial compass enabling maintenance of preferred bearings perpendicular to the home shore in *Goniopsis cruentata* (Schöne, 1963), *Uca tangeri* (Altevogt and von Hagen, 1964), and *U. pugilator* (Herrnkind, 1966, 1968*a,b;* the orientation of fiddler crabs to be treated in detail later). This mechanism also guides the essentially littoral form *Hemigrapsus oregonensis* parallel to shore in apparent shelter-seeking movements (van Tets, 1956, unpublished thesis) and possibly directs individuals of *Ocypode ceratophthalma* to a specific burrow site (Daumer *et al.*, 1963). As in amphipods, isopods, and wolf spiders, both the sun and polarized light from the zenithal region serve as guidance cues either separately or in conjunction (Altevogt and von Hagen, 1964; Schöne, 1963; Daumer *et al.*, 1963; Herrnkind, 1968*a,b*). In fact, sensory physiological and behavioral experiments on these and other crustaceans, such as *Podophthalmus, Cardisoma* (Waterman, 1966; Waterman and Horch, 1966), and *Carcinus* (Shaw, 1966, 1969; Horridge, 1966*a*, 1967), suggest that sun positional fixation and polarized-light perception are common capabilities in crustaceans as well as in insects and arachnids (Waterman, 1966; Waterman and Horch, 1966; von Frisch, 1967).

A second visual mechanism less well studied is orientation to landmarks. For example, *Ocypode saratan* females orient to the sand cones constructed by the males (Linsenmair, 1965, 1967). The cone owners also demonstrate the ability to gauge the distance from their burrow opening to the cone. Hughes (1966) reports that *Ocypode ceratophthalma* are caused to misjudge the position of their burrow opening by repositioning prominent objects near the burrow. Some fiddler crabs demonstrate a spontaneous orientation toward natural landmarks, and others are capable of learning to approach objects under artificial conditions (Herrnkind, 1965, 1968*a,b;* Altevogt, 1965). These results suggest responsiveness to certain areas of optical contrast, more advanced form vision in recognition of definitive shapes, and, perhaps, recognition of spatial configurations of multiple objects. Thus far there are few analytical studies of crustacean form vision and related processes (Williamson, 1951*b;* Salmon and Stout, 1962; Horridge, 1966*b;* Korte, 1966), although preliminary results suggest that fiddler crabs can distinguish basic shapes such as triangles and circles of equal area (Langdon and Herrnkind, unpublished data).

Nonvisual mechanisms, especially kinesthesis and gravity perception, are important to certain oriented movements. Kinesthesis is exhibited by both *Ocypode* and *Uca* during return to burrows from short distances (von Hagen, 1967, and see p. 13). Proprioceptive senses, in general, operate in integration of oriented locomotion by permitting crabs to change their horizontal body axis while maintaining a directed course and by compensating eyestalk angle to maintain a stable visual field when moving over uneven surfaces (Cohen and Dijkgraaf, 1961).

2. Establishment of Preferred Bearings

Processes establishing preferred directional responses in respect to the celestial compass have been heretofore examined only cursorily (Herrnkind, 1968*a*). However, such preferences are assumed to be learned, since populations of decapods on any shoreline are derived from planktonic larvae from scattered parent populations often living on beaches with different compass bearings (Schöne, 1963; Altevogt and von Hagen, 1964). The demonstrated capacity of directional learning by *Ocypode ceratophthalma* (Daumer *et al.*, 1963), *Uca tangeri* (Altevogt, 1965), and *U. pugilator* (Herrnkind, 1966, 1968*a*, *b*) supports this contention. Young *O. ceratophthalma*, having moved 50 cm from an artificial burrow in a circular chamber (permitting only a view of polarized sky light), dart back in the opposite compass direction when startled by rapid horizontal rotation of the chamber (Daumer *et al.*, 1963). Kinesthetic and landmark cues are, of course, obviated by the rotation, indicating that the crabs "remember" the route on a basis of polarization cues. Mature fiddler crabs adopt compass preferences in similar situations, as discussed in detail on pp. 15–18. Crustacean learning is also demonstrated in other situations (Cowles, 1908; Schwartz and Safir, 1915; Altevogt, 1964, 1965; Datta *et al.*, 1960; Schöne, 1961, 1965), and, while all these experiments involve artificial conditions, they do demonstrate the widespread capability of learning in the group.

Despite the demonstrated learning ability, an understanding of the contribution of genetic, maturational, and/or experiential processes to orientation requires examination of development of orientational behavior during ontogenesis under natural conditions, as well as analysis of the underlying mechanisms under controlled conditions. The fruitfulness of this approach is apparent in the discussions of amphipod and wolf spider orientation and is necessary for comparisons between these groups and decapods. A preliminary study of orientation during ontogeny in *Uca pugilator*, undertaken to fill this gap, is reported here along with a reasonably extensive review of orientation in fiddler crabs.

III. ORIENTATION BY FIDDLER CRABS, GENUS *UCA*

A. Orientation Mechanisms and Adaptiveness

1. Escape Responses

The most conspicuous and characteristic oriented movements by adult fiddler crabs are escape responses induced by certain visual and substrate-borne acoustic stimuli. These escape movements are of two types in the species studied: (1) rapid direct homing to individual burrows from distances of up to 1 m (Altevogt and von Hagen, 1964; von Hagen, 1967; (2) locomotion, with more variable heading, generally from the lower beach landward to the burrow areas or supralittoral zone (Altevogt and von Hagen, 1964; Herrnkind. 1965, 1968*a*,*b*). The former, near-orientation (from Nahorientierung, after Altevogt and von Hagen, 1964), is hole-directed, independent of the crab's position relative to it, while the latter, far-orientation (Fernorientierung), is directed landward, independent of the position of the burrows relative to the crab.

The two responses are independent and are not known to combine to direct a fleeing crab to its burrow over long distances. This independence results from the operation of different orientational mechanisms in each. The near-orientation is guided kinesthetically through proprioceptive cues, while the far-orientation is primarily visually directed by celestial cues and landmarks. This being known, the behavioral characteristics, adaptive function, and operation of these responses in nature can be understood.

2. Near-Orientation

The near-orientation is exhibited by crabs actually occupying a specific burrow. Most crabs in a population show such an affinity when they initially emerge at ebb tide and when they obtain an unoccupied burrow, or build a new one, just prior to high tide. Other individuals, particularly waving males, are territorial and may hold a single burrow for several days (Crane, 1958). The crabs typically emerge from the burrow and move radially outward, depositing feeding pellets. At 20–50 cm, they stop and periodically return directly to the burrow (either spontaneously, to wet the gills, or if startled), emerge again, and move off radially in another direction until eventually a spoke-shaped pattern is formed by their paths and rows of feeding pellets. They dart unhesitatingly from any position to their own burrow even though other abandoned holes may lie nearer to their return path.

The orientation is not dependent on visual guideposts, since the crabs unerringly return to the burrow even on dark nights and when visual cues are modified (Altevogt and von Hagen, 1964). However, crabs removed bodily from the substrate and placed within 50–60 cm of their burrow do not return directly to it; rather, they wander about crisscrossing the area, some relocating the burrow. Crabs displaced beyond this distance are generally unable to relocate the original burrow and may attempt to enter others or begin to dig a new one. Removing the crab from its position on the substrate deprives it of any kinesthetic course reference relative to the burrow and causes disorientation.

The kinesthetic mechanism in *Uca rapax*, and probably the other members of the genus, controls both the direction and distance to the burrow (von Hagen, 1967). Thus, if a crab is caused to miss the burrow, it stops at the approximately correct distance and then makes zigzag searching movements about the immediate area. The mechanism apparently also permits integration of short detours in the return path, as suggested by the bends in some rays of the pellet pattern.

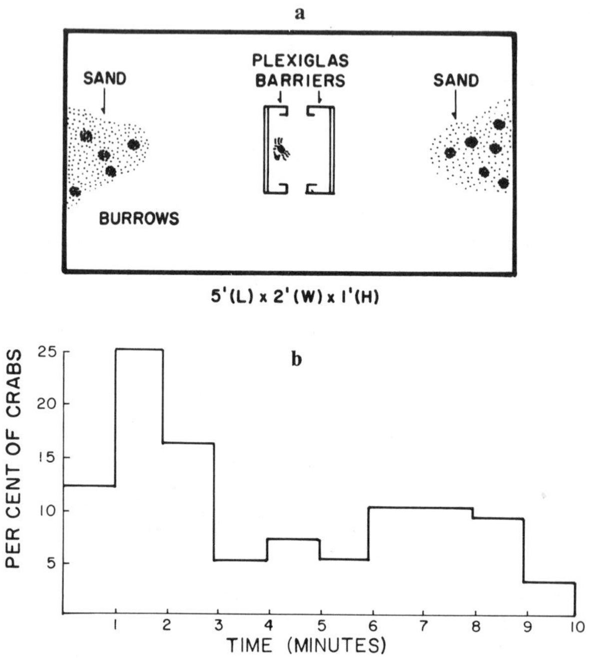

Fig. 1. (**a**) Detour test apparatus for sand fiddler crabs, *Uca pugilator*. Freshly collected crabs are permitted to establish burrows in damp sand at either end of the chamber and otherwise acclimate for 24 hr. Thereafter, a crab removed from a burrow and placed on the bare chamber floor runs directly to the sand. Plexiglas barriers with a clear face 15 cm long and opaque lateral projections 5 cm long are placed in the center as shown (not to scale). A crab released from the center must now circumnavigate the barrier to reach the sand. (**b**) Successful detour performance of 68 crabs (of 140) within 10 min during the initial trial. In all cases, the crabs first ran toward the sand and collided with the plexiglas but eventually reversed course and circumnavigated the barrier in a smooth movement, without erratic collisions and without maintaining contact with the barrier.

The presence of detour integration is partially substantiated by the detour ability of *Uca pugilator* in a slightly different situation (Herrnkind, 1965). The crabs, in this case, can initially see their sand burrow area from the center of an arena through the glass wall of a barrier (Fig. 1a). To reach the burrows, a crab must move away from the goal and around the opaque side-walls. Crabs first run directly at the sand, collide with the glass, and climb against it. Individuals detouring the barrier do so in a smooth continuous movement, without random running and without maintaining contact with the barrier (Fig. 1b). The crabs lose sight of the goal while circumnavigating the opaque portions of the barrier, but this usually does not cause a change in the smooth detour movement. Fiddlers, like many other homing animals, can change or even reverse their course to reach a goal.

The adaptive significance of the kinesthetic near-orientation lies in its high precision of directing crabs to their own burrow as rapidly as possible. A crab is not confused by the myriad of visual stimuli around it and does not have to visually select its burrow from all the others. In fact, the stalked compound eye cannot readily view nearby objects on the substrate below (Herrnkind, 1968*a*). Finally, upon reaching the opening, the crab is assured of the proper aperture and an unoccupied burrow. This is particularly important in evading predatory birds that selectively attack those crabs protruding from an undersized hole or those prevented from entering by an established resident.

3. Far-Orientation

The far-orientation is more complex and subject to greater variability than the near-orientation. Generally, crabs feeding on the lower beach move landward when disturbed. These crabs do not hold a particular burrow and may be up to 50 m from the burrow region. If the disturbance is intense or persistent, they continue upshore until reaching the burrow area, where they enter any unoccupied burrow in their path, or the supralittoral vegetation, where they take refuge in suitable interstices (Herrnkind, 1968*a*,*b*). Landward movement is demonstrably more advantageous than seaward movement in most habitats, since aquatic predators such as portunid crabs are numerous and readily attack submerged fiddlers. In fact, very strong landward orientation is induced by displacing crabs offshore, in which case they may intersperse running with burying in the substrate for periods of up to several minutes.

The far-orientation is directed primarily by visual mechanisms relying upon celestial cues (the sun or polarized sky light from the zenithal region or both) and gross landmarks, such as mangroves or clumps of grass on the upper shore (Altevogt and von Hagen, 1964; Altevogt, 1965; Herrnkind, 1965, 1968*a*,*b*). The celestial orientation is demonstrated by the tendency of crabs to gather in the appropriate landward compass reference in a circular chamber (Figs. 2 and 3) permitting a view of the sky and sun, polarized zenithal region, or a mirror image of the sun (Altevogt and von Hagen, 1964; Herrnkind, 1968*a*,*b*). Since the crabs usually climb the chamber wall for up to 20 min and continue to move landward when released from the chamber onto the beach, their behavior is interpreted as a manifestation of the escape response. The celestial mechanism is apparently time compensating, since crabs orient by celestial cues at all daylight hours. Orientation by celestial cues involves simultaneous and synergistic operation of separable visual elements responding either to the sun or polarized light (Waterman and Horch, 1966). In *U. tangeri*, only that polarized light entering the apical ommatidia causes an oriented response (Korte, 1966). Celestial orientation is especially

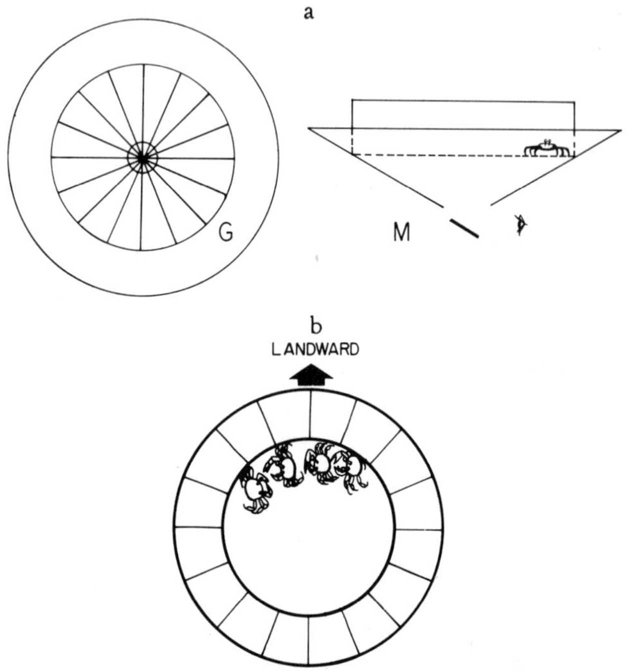

Fig. 2. (**a**) Schematic top and side views of 30 cm diameter, clear plastic test pan and removable sloping base. Test crabs were placed, either singly or in groups of three to ten, in the pan and observed from below directly or by mirror (M). Positions of crabs, recorded at intervals of 1–5 min, were determined from the polar coordinate grid lines (G) every 22.5° on the pan floor. (**b**). Illustration of posture and positions taken by adult sand fiddler crabs in the test chamber in response to guidance cues such as the sun or polarized sky light or both (diagram not to scale).

important in permitting crabs to direct and maintain a course when travelling underwater or through tall grass or debris.

Orientation to landmarks resembles a positive telotaxis (Jander, 1965); the crabs travel directly toward an object or region of optical contrast as a "goal" (Herrnkind, 1965, 1968*a,b;* Altevogt, 1965). This differs from the celestial menotaxes, where the crabs orient at some appropriate angle to the sun or plane of polarization whether or not a burrow area "goal" is present. The landmark orientation is strongest in the absence of celestial cues, at which time sand fiddlers orient to real or artificial landmarks (black or white vinyl screens) independent of the appropriate landward bearing (Herrnkind, 1968*a,b*). In nature, the landmark object (mangrove, grass clump, etc.) serves either as a guidepost or a source of cover. However, conspicuous landmarks often lie landward, so escape responses directed at them, like celestially guided movement, generally carry the crabs through burrow regions where many obtain refuge. Specific landmark objects may also aid crabs in reorienting to a burrow or to a specific feeding site, as suggested by Altevogt's (1965) study.

Certain natural situations occur such that landmarks lie in some opposing relationship to the landward direction which is indicated by celestial cues, e.g., mangroves growing offshore or on the other side of a tidal canal. Hence,

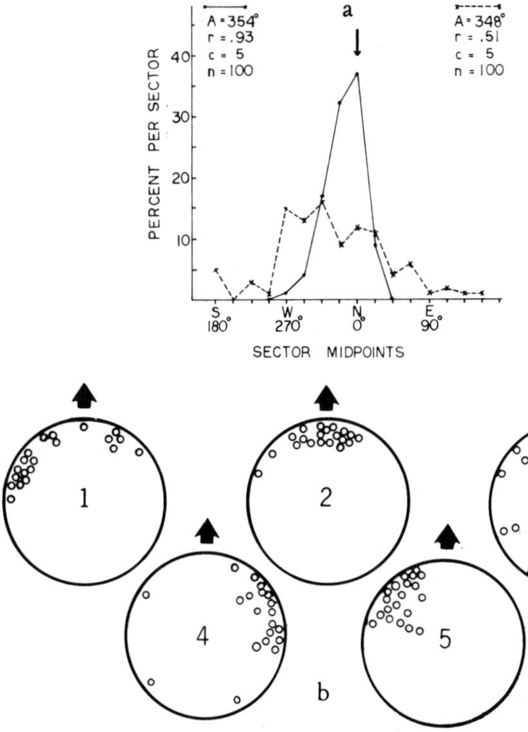

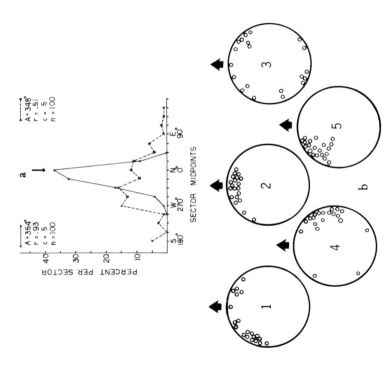

Fig. 3. **(a)** Graphs representing 360° circular distributions plotted on a linear scale over the compass range (S = sourth, N = north, E = east, W = west). The percent of points lying within each of the 16 22.5° sectors is plotted at each sector midpoint. *A* is the mean angle; *r*, the mean vector; *c*, the number of test crabs; *n*, the number of positions recorded. The vertical arrow (↓) represents the landward compass bearing for the site of collection of the experimental animals. The solid line shows the distribution for five crabs tested as a group, while the dashed line shows the distribution for the same crabs tested individually. The larger dispersion of solitary crabs is indicated by the lower mean vector value. ($r = 0.51 < 0.93$; a value of 1.0 indicates the highest concentration, i.e., all points at a single locus, while 0.00 indicates a perfectly uniform distribution. Refer to Appendix Table I and see Batschelet, 1965, for details.) **(b)** Scatter diagrams of positions recorded for the five crabs tested individually. Note that the weak orientation of crabs 1 and 3, as well as the deviation from landward (black arrow) by 4 and 5, is eliminated by testing the crabs together as shown in (a).

the crabs cannot travel toward the landmark and orient landward at the same time. The choice made by sand fiddlers under these conditions varies, depending upon several factors (Altevogt, 1965; Herrnkind 1965, 1968*a*,*b*): (1) most commonly, a crab orients landward for some distance, then, if pursued further, veers toward the nearest object; (2) if a crab on a celestial bearing becomes desiccated, it usually shifts to orientation to a landmark; (3) some proportion of crabs in a population initially set a course directly to a landmark even under a clear sky (this may reflect the absence of a landward response in these individuals, as observed also by Altevogt in *U. tangeri*); (4) landmark orientation often appears in crabs displaced to novel areas where landmarks and celestial cues lie in a different relationship than in the familiar part of the beach; (5) crabs moving in the inner portion of droves apparently orient themselves according to their neighbors, rather than all individuals orienting independently by landmarks or celestial cues. I believe this social effect gives rise to the grouping (Altevogt and von Hagen, 1964) observed in test chambers, where crabs tested in a group exhibit a more compact distribution about the landward reference than when each crab is tested individually (Fig. 3). Grouping may actually represent a pooling of directional information, the well-oriented crabs being followed by those with weaker preferences.

In far-orientation, the transfer between landmark and celestial guidance is apparently adaptive, since neither guidepost alone always provides directional information appropriate to every situation. A shift from landward movement to orientation to an object can facilitate evasion of a predator or lead a crab to an area of decreased insolation and increased humidity. The ability to orient by any or all of the three guidance cues (sun, polarized light, landmarks) suggests that crabs in natural situations are seldom disoriented.

B. Factors Controlling Preferred Bearings

1. Habitat Characteristics

The directional preference of adult fiddlers in any population depends primarily upon the geographic configuration and topography of the shoreline, as well as the stability of the shore–water spatial relationship. Thus, crabs living on relatively straight beaches exhibit compact distributions about the landward bearing during celestial orientation (Fig. 4a; Herrnkind, 1968*a*). Crabs from shores bordering sinuous tidal canals show greater dispersion (Fig. 4b), while those from tidal swamps having no permanent land–water relationship often exhibit a random or nearly uniform scatter (Fig. 4c). This behavior is also characteristic of wolf spiders in analogous situations, as described previously (Papi and Tongiorgi, 1963).

Other ecological factors also affect orientation behavior, both in regard to directional preferences and in response to disturbances. For example, the shoreline typical for *Uca pugilator* and *Uca rapax* is a beach bordered landward by vegetation and seaward by shallows. However, along some tidal canals the landward area of dredged fill is dry and devoid of vegetation while the edge of the bank drops nearly vertically to a depth of several meters. Here, fleeing crabs run seaward, climb down the bank into the water for several centimeters, and enter any available crevice. This response seems adaptive, even though the crabs are exposed to aquatic predators, since the bank edge provides the only refuge aside from the burrows. Another exception to the landward preference is observed on very broad beaches, or flat spits, where crabs wander in droves 30 m or more from the burrow areas. Here, disturbances cause the drove either to move away as a group, disperse from the point of disturbance, or scatter in various directions. Although these responses do not lead directly to refuge, it is also unlikely that the crabs could reach the burrow area before being overtaken by a predator. In summary, the behavior patterns characteristic of individuals from different beaches vary according to the ecological conditions present.

2. Endogenous Components

The correlative behavioral differences in orientation by crabs in various habitats are the result of internal differences and not just the effects of immediate external stimulation in each environmental situation. This is clearly shown by the differences in directional preference (in response to celestial cues) of crabs from various shorelines when tested in the same circular chamber. Understanding how particular reponses arise in each situation requires understanding of the operation of the two main interacting components: (1) the biological clock mechanism and (2) the process(es) responsible for association of a directional preference with celestial cues.

3. Temporal Features

Knowledge of biological clock operation in fiddler crabs as it integrates orientation to celestial cues is inadequate despite the voluminous research on activity, metabolism, and chromatophore rhythms (Brown, 1961; Barnwell, 1968). The orientation clock in *Uca* should be analyzed in similar fashion to, and to the same extent as, that in other arthropods (von Frisch, 1967) and fishes (Hasler, 1967), since the available information only indicates its role and operational characteristics. For example, sand fiddlers dug from burrows at high tide orient landward in a test chamber just as those at low tide do (Herrnkind, 1968*a*). Similarly, crabs repeat their directional preference at various times during daylight hours, whether held indoors between

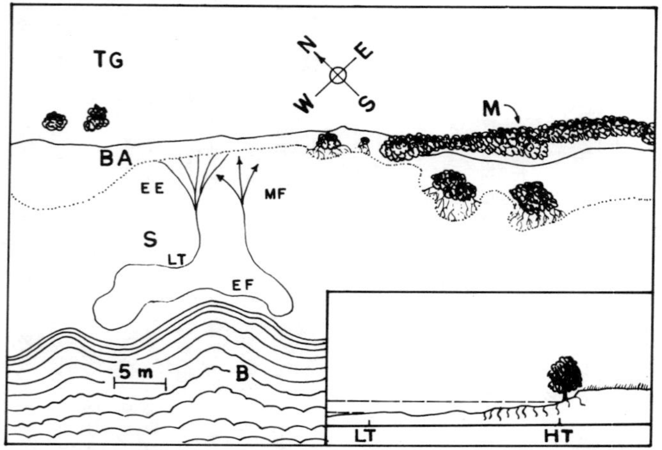
TG
N
E
W
S
M
BA
EE
MF
S
LT
EF
5 m
B
LT
HT

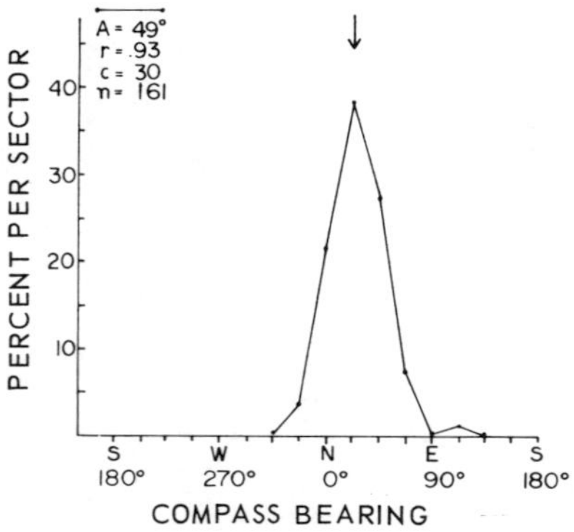
A= 49°
r= .93
c= 30
n= 161
PERCENT PER SECTOR
40
30
20
10
S
180°
W
270°
N
0°
E
90°
S
180°
COMPASS BEARING

a

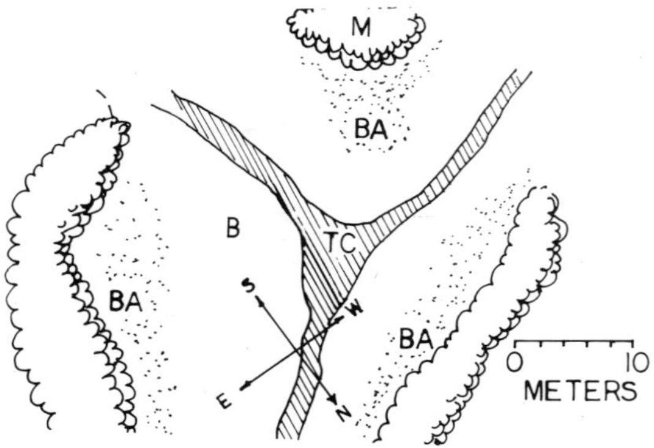
M
BA
B
TC
BA
BA
S
W
E
N
0
10
METERS

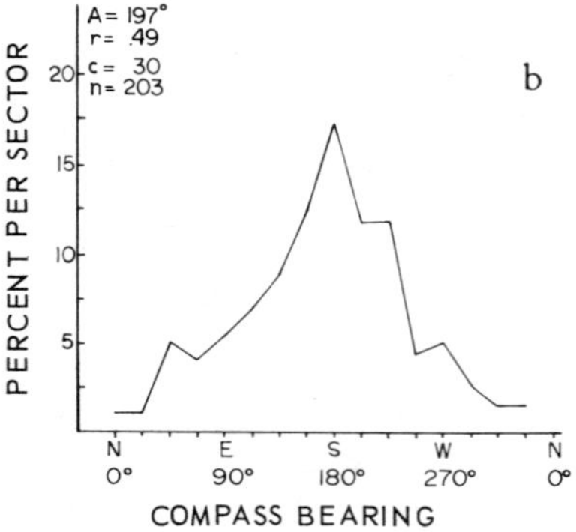
A= 197°
r= .49
c= 30
n= 203
PERCENT PER SECTOR
20
15
10
5
N
0°
E
90°
S
180°
W
270°
N
0°
COMPASS BEARING

b

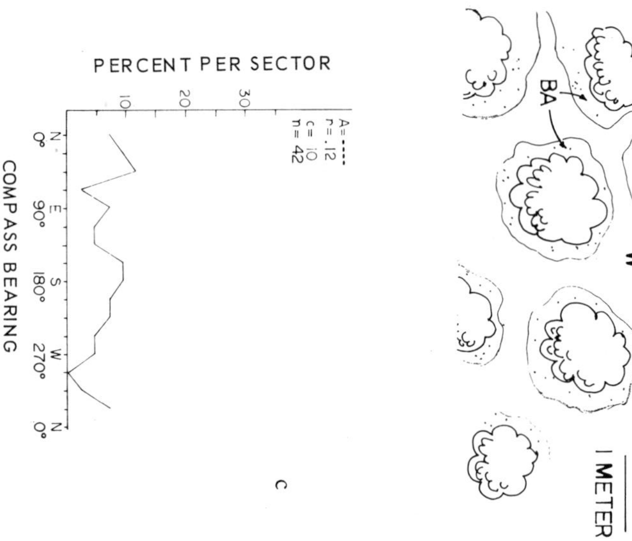

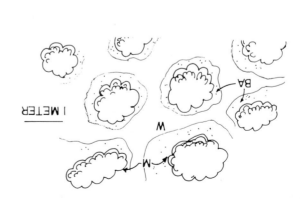

Fig. 4. (a) Top: Diagram of a straight shoreline habitat. The level sand beach is exposed to a width of 20–40 m at low tide (LT). It is bordered to the northeast by mangroves (M) and a sharply sloping bank (see inset) covered with shore and terrestrial grasses (TG) and bordered to the southwest by shallow water (B) and *Thalassia* beds. Sand fiddler crabs, *Uca pugilator*, inhabit extensive burrow areas (BA) along the upper shore (HT) and wander in large herdlike groups, or droves, over the limits of the exposed beach at low tide as shown (arrows). Crabs emerge and congregate at early ebb tide (EE), migrating to the lower beach and feeding during low tide (LT) and early flood (EF). By midflood, the drove moves landward and disperses, individuals locating or building a burrow before high tide. Bottom: Directional preference of crabs living on this shoreline showing a strong landward component (↓) in response to celestial cues. **(b)** Top: Diagram of a region adjacent to a sinuous tidal canal (TC). M, mangroves; BA, burrow areas; B, beach. Bottom: Distribution by a group of 30 crabs from this area in response to celestial cues. **(c)** Top: Diagram of a mangrove complex without definite shore–water references. W, water; BA, burrow areas. Bottom: A scattered distribution, in response to celestial cues, by crabs from this latter region.

tests or in view of the sky. Apparently, they do not have to reset their directional reference, relative to celestial cues, when they emerge after 6–18 hr below ground over a tidal cycle or overnight.

A tidal influence on directional tendency, superimposed on the above solar day compensation, appears in crabs left undisturbed underwater in celestial test chambers and monitored for several days (Table I). The crabs group about the landward bearing and remain there until the time of low tide at the collection site. They then disperse and regroup within an hour in the seaward bearing, where they remain for an hour. Thereafter, they move progressively landward, attaining that bearing just prior to peak high tide (Fig. 5). This cycle is repeated for up to 2 days, after which the directional preferences weaken.

I tentatively interpret the shift from landward to seaward at low tide and back to landward at flood tide as a manifestation of the normal activity pattern with appropriate orientation to the habitat. In fact, other members of the same population, in the field, moved seaward to feed on the lower beach and returned upshore within at most 2 hr of the test crabs (Fig. 4a). The clock, then, not only compensates for temporal changes in position of celestial

Table I. Fluctuation of Preferred Bearings and Dispersion with Tide[a]

High tide		Low tide	
A	*r*	*A*	*r*
23°	0.78	212°	0.54
16°	0.76	29°	0.41
38°	0.64	202°	0.22
33°	0.73	32°	0.65
29°	0.81	53°	0.52
28°	0.99	46°	0.51
19°	0.99	26°	0.35
26°	0.76	29°	0.46
32°	0.88	37°	0.58
28°	0.69	202°	0.30
20°	0.60	188°	0.29
32°	0.65	2°	0.55
20°	0.60		
22°	0.99	Avg. deviation from landward = 67°	
19°	0.99		
33°	0.79		Avg. *r* = 0.45
Avg. deviation from landward = 6°			
	Avg. *r* = 0.79		

[a] Possible tidal fluctuation in preferred bearings (A = angle in degrees) and dispersion (r = mean vector) by 72 adult fiddler crabs, *Uca pugilator*, held under 10 cm of water in celestial test chambers (ten crabs in each) over two 2-day periods. Landward, 23° (NNE); seaward, 203° (SSW). Values of r approaching 1.00 indicate compact grouping, while decreasing values indicate more dispersion; an r value of 0.00 indicates uniform distribution (Batschelet, 1965).

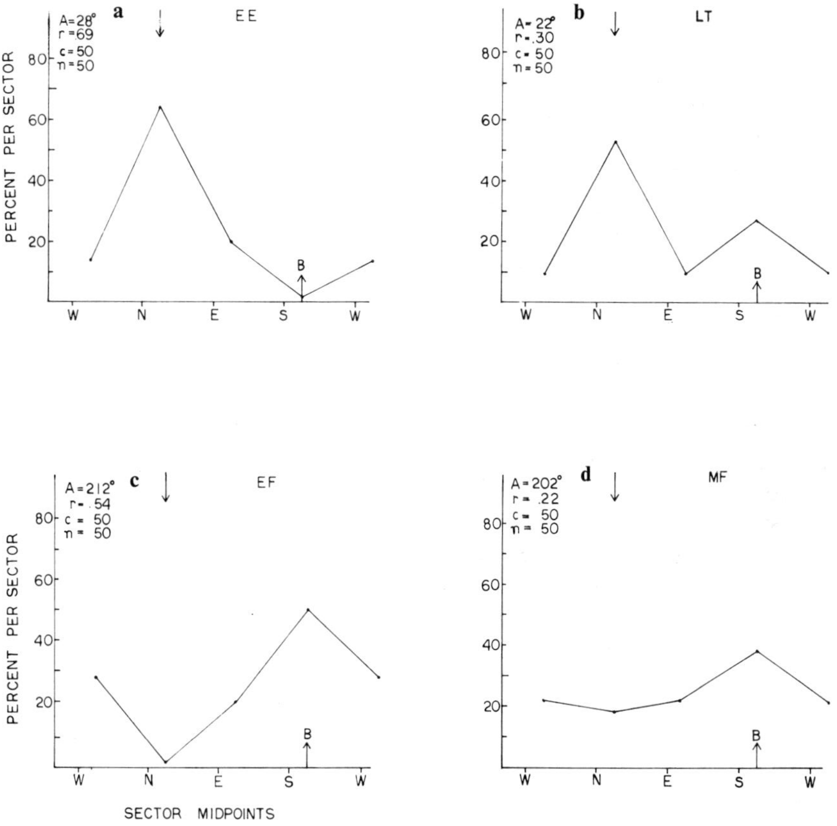

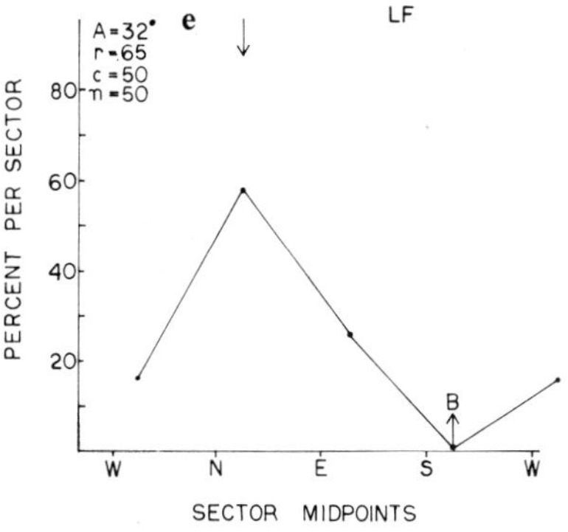

Fig. 5. Selected sequence of distributions (combined) exhibited by five groups of ten crabs in separate pans over a tidal cycle from late ebb to late flood. **(a)** At the beginning of the ebb tide (EE), a landward preference is apparent. B represents the seaward bearing. **(b)** One hour later, during the low tide period (LT), some crabs orient seaward (B). **(c)** At 2 hr, or early flood (EF), the majority orient seaward. **(d)** At midflood (MF) the seaward preference weakens. **(e)** At late flood (LF) the group again orients landward. Concurrent field observations revealed a parallel sequence of movements by crabs at the collection site (Fig. 4a) preceding the activities of the test group by 1–2 hr.

cues but may also control directional components of cyclic activities. Further studies are needed to determine the validity of this hypothesis.

4. Role of Experience

Preferred escape bearings may vary during a crab's lifetime. For example, individuals and groups of crabs are known to migrate along the beach (Altevogt, 1957) to places where the compass bearing of the shoreline may be different from that of the previously occupied area. A stereotyped directional preference implies constant disorientation in the new situation, but this is not observed. Similarly, storm erosion often restructures parts of a shoreline, but this does not result in observed disorientation of resident crabs. Finally, the directional preference exhibited at any time is labile and wanes in crabs held indoors, or in barren containers, after 1–4 days (Altevogt and von Hagen, 1964; Herrnkind, 1968*a*). From this, one would expect hibernating crabs (in temperate areas) and those forced to stay underground during extensive storms to lose their directional preference and have to re-establish it upon emergence.

Experimental results demonstrate that a particular bearing can be induced in simulated habitats consisting of a shore and shallow water provided with a guidance cue, such as overhead polarized light. Crabs under these conditions eventually adopt a preference to orient to the polarization plane relatively parallel to the shore–water bearing of their habitat (Herrnkind, 1968*a,b*, and see pp. 33–38). Furthermore, Altevogt (1965) reports training individuals of *U. tangeri*, having no apparent spontaneous preferred direction by celestial orientation, to rapidly adopt one by forcing them to repeatedly run about 50 cm to a simulated burrow at a fixed compass reference from the release point. The crabs establish orientation toward a landmark, independent of celestial cues, in the same way. Directional preferences appear after an average of 20 such runs and are repeated on subsequent trials the following day. Crabs demonstrating strong landward preferences, however, are not readily trainable by this technique, suggesting that crabs normally learn a new bearing only after the previous one wanes. This may occur in nature during long periods underground, as previously noted, and during cyclic internal changes manifested as shifts in activity patterns over periods of days or weeks, e.g., from feeding with the drove to aggressive wandering in the burrow area, or waving and territorial behavior (Crane, 1958; von Hagen, 1962). Another possible cause for waning of a directional preference is extinction of the response in the absence of reinforcing conditions. As suggested by Schöne (1965) and Lorenz (1965), the demonstrably important process of establishment of directional preferences in *Uca* and other homing and migrating species, such as *Goniopsis cruentata* (Schöne, 1963), *Ocypode* *ceratophthalma* (Daumer *et al.*, 1963), *Panulirus argus* (Schöne, 1965; Herrnkind, 1969), and *Pagurus longicarpus* (Rebach, 1968), seems to be a fruitful area for investigation of associated learning phenomena in crustaceans.

C. Orientation During Ontogeny in *Uca pugilator*, the Sand Fiddler Crab

1. Rationale for Investigation

Clearly, the orientational behavior of adult fiddler crabs can be modified through experience as environmental situations change. Nevertheless, one cannot interpret the contribution of genetic components or ontogenesis to the behavior by studying only mature subjects. Since the developmental processes in semiterrestrial Brachyura are greatly different from those of shore-living amphipods, isopods, and arctosid spiders, the ontogeny of orientation in *Uca* cannot be assumed analogous to that of the latter groups. Moreover, the transitional ontogenetic stages in fiddler crabs differ morphologically, physiologically, ecologically, and behaviorally from the adult and, successively, from one another (Table II). Therefore, their orientational repertoire cannot be assumed equivalent to that of the adult, In fact, Altevogt and von Hagen (1964) describe several ontogenetic differences in orientation of *U. tangeri*; i.e., early stages, 2.6–5.1 mm in carapace width, do not spontaneously orient by polarized light, and crabs 7–15 mm in carapace width exhibit seaward preferred directions, while those over 15 mm prefer the landward bearing. An understanding of the function and operation of orientation in fiddlers requires further knowledge of ontogenetic aspects.

An initial analysis of ontogenesis of visual orientation in the sand fiddler crab (*Uca pugilator*), from first instar to adult, is described in the following. The research aims were to characterize orientational responses to visual stimuli during development, to relate orientational performance to ecological situation, and to determine the genetic, maturational, and experiential factors contributing to the orientation.

2. Test Animals and Methods

Test crabs were derived from two sources. Wild crabs, from first instar to adult, were collected on Virginia Key, Miami, Florida, from a sand beach characterized by a relatively straight shoreline, sloping gradually seaward into shallows, and bordered to landward by mangroves and beach grass (Moore *et al.*, 1968, discuss the ecology of this area in detail). Adults from this site demonstrate strong landward orientation to the north-northeast (NNE = 22.5°) in response to celestial cues (Fig. 4a). Also, young crabs were laboratory reared from the eggs of fertilized females collected at this

Table II. Summary of Organismic Characteristics, Behavioral Ecology, and Orientation of Sand Fiddlers, *Uca pugilator*, During Ontogeny

Age and stage	Organismic characteristics	Behavioral ecology	Orientation
Under 1 week (stage 1–3)	One to two millimeters in carapace width; locomotion feeble, posterior pair of pereiopods adapted to clinging; low resistance to desiccation; camouflage coloration.	Remain in wet part of beach, in tide pools, and clinging to *Spartina* grass or debris; occasional movement over open wet sand.	Strong telotactic orientation (attraction) to objects of optical contrast; tend to move into lighted areas if moisture present.
1 week to 2 months (stages 3–8)	Two to four millimeters in carapace width; all pereiopods assist locomotion; unequal chelae appear in males; stalked eyes held more vertical; camouflage coloration.	Dig burrows over wide, wet zone between midtide level and high tide level; occasionally reside in abandoned burrows of older crabs; move about more than previous stages but still restricted to wet areas.	As above.
2–6 months (stages 8–12)	Four to six millimeters in carapace width; male chela large and stout; increased locomotor rate; increased resistance to desiccation.	Reside among older crabs in colonies near high tide region; agonistic behavior appears; generally little interaction with older conspecifics; movements mostly in burrow area and on wet part of beach.	Telotaxis remains but some crabs show landward directional preferences to celestial cues.
6–10 months (stages 12–15)	Six to nine millimeters in carapace width; can tolerate much longer periods of exposure than previous stages; adult color pattern appears.	Reside in main burrow area at high tide level with adults; agonistic behavior; rudimentary waving appears in older individuals; these latter also join in wandering droves as do full adults.	Telotaxis appears mostly when celestial cues absent; strong celestial orientation to landward during escape response in older crabs.
10 months through adulthood (stages 15–?)	Ten to twenty-five millimeters in carapace width; sexual maturity; ornate color pattern on carapace.	Full social behavior; waving males and ovigerous females reside above high tide region; drove wandering prevalent.	Peak of landward-directed celestial orientation; landmark orientation in absence of celestial cues, under desiccating conditions, and when closely pursued.

beach (or from females held and bred in the laboratory). The larvae and early instars were raised completely indoors (details in Herrnkind, 1968*c*) under diffuse artificial light and were not exposed to a simulated shore or the sun. Responses of both groups were tested, as will be described, and compared to determine the influence of environmental experience, obtained by the wild crabs, on the basic spontaneous repertoire demonstrated by naive, laboratory-reared crabs.

Individuals or groups of up to ten were placed in 30 cm diameter pans with transparent bottoms and allowed to acclimate for 2 min. Thereafter, the positions of stationary crabs were recorded, from below, every 15–30 sec for 5–10 min. Four experimental situations were investigated: (1) control tests were conducted in a pan with a transparent bottom and the vertical sides sprayed with dull white paint (an overhead 60 W light with reflector and white paper diffuser provided a uniform white light field to the sides and above); (2) tests for the responses to regions of optical contrast (simulated landmarks) incorporated a black screen, covering a 45° arc, in the white-walled chamber (with the white paper diffuser and overhead light); (3) celestial orientation was tested in a similar transparent pan placed in a uniformly white, sloped base to eliminate landmarks from the visual field (Tests were conducted outdoors on clear days from 0900 to 1100 hr and 1400 to 1600 hr, purposely avoiding the zenithal solar altitude which is nearly overhead in summer at 25°N latitude); (4) the relative influence of celestial cues and screen was determined by presenting the black screen 180° from the preferred compass bearing exhibited by crabs in (3) (where no preferred direction was shown, the direction exhibited by adults (NNE) was used as the celestial reference). Seawater up to the level of the carapace was provided to prevent desiccation in all experiments unless otherwise stated. Three size–age groupings were tested in each of the four situations: 1.5–3.0 mm carapace width, 1 day to 2 months old; 3.5–6.0 mm, 2–5 months old; 6.5–8.0 mm, 5–8 months old. Crabs attained reproductive age when 9–11 months old. The size–age relationships were determined from growth rates of 100 crabs raised in simulated habitats (Herrnkind, 1968*a*,*c*).

3. Results and Interpretation

a. Control. The control groups did not exhibit preferred directions in the uniform light field (Fig. 6), thus eliminating the possibility of extraneous visual cues in the test chamber and indicating that pervasive factors, such as magnetism or electric fields, were insufficient to induce orientation under the test conditions. The scatter of data from control crabs resulted largely from their constant locomotory activity about the chamber. Moreover, individuals moved across the entire bottom of the chamber, rather than remaining along the perimeter as do crabs presented guidance cues. This behavioral difference

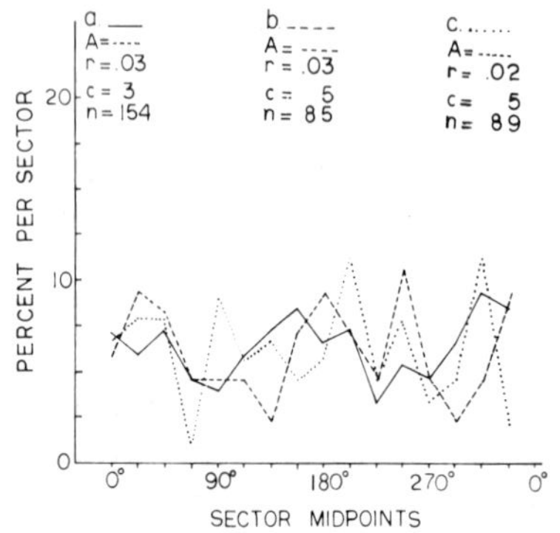

Fig. 6. Distributions of young, wild sand fiddlers in an evenly illuminated circular test chamber. **(a)** Under 2 months old **(b)** 2–5 months old; **(c)** 5–8 months old.

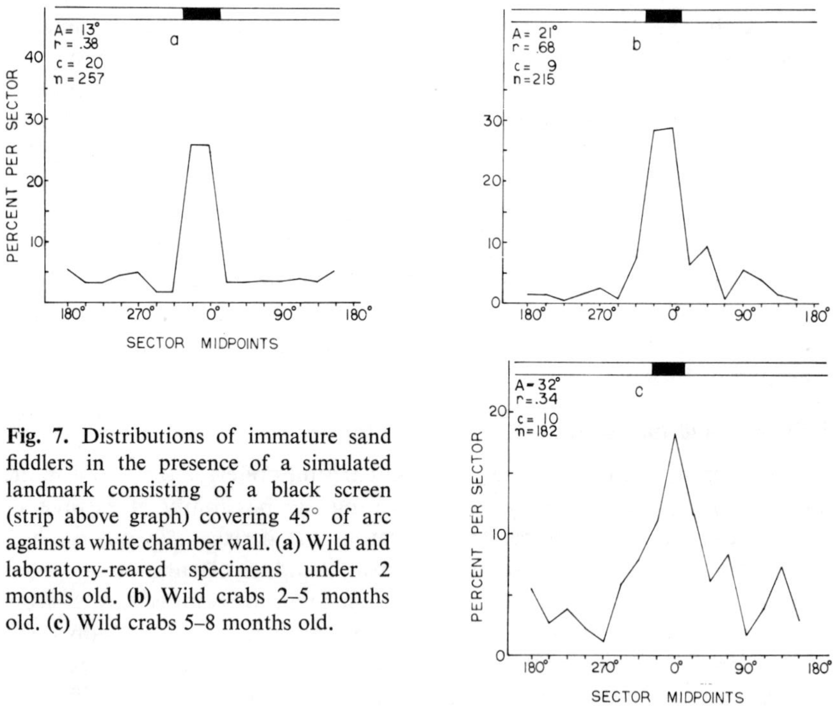

Fig. 7. Distributions of immature sand fiddlers in the presence of a simulated landmark consisting of a black screen (strip above graph) covering 45° of arc against a white chamber wall. **(a)** Wild and laboratory-reared specimens under 2 months old. **(b)** Wild crabs 2–5 months old. **(c)** Wild crabs 5–8 months old.

provided a qualitative indication of orientation, in addition to the quantitative measure of preferred directions and degree of dispersion (Batschelet, 1965).

b. Landmarks. All test groups exhibited maximum concentration in the sectors adjacent to the black screen, although the younger specimens (less than 5 months) responded more strongly than the older (Fig. 7). Crabs under 2 months old, both wild and laboratory-reared, travelled directly to the screen and usually remained adjacent to it, or within a few millimeters of its side edges, for periods of several minutes. This response was observed in crabs up to 3–4 months old, after which the spontaneous attraction to the screen noticeably declined. The equally strong orientation by wild and laboratory-reared specimens suggests that the spontaneous response to regions of optical contrast is not markedly influenced by environmental experience for the first few months, at least not in the habitat situation tested.

The attraction of early instars to optically contrasting objects correlates well with the behavior of these stages in the enviornment. The first few instars do not dig burrows; rather, they seek shelter under pieces of debris, in crevices of driftwood, or on the lower portion of *Spartina* plants (Hyman, 1922; Herrnkind, 1968*a*,*c*). The initial orientation to such objects may be visual, as suggested by the strong, positive telotaxis by young test crabs. This mechanism is particularly adaptive, since early instars in the open are very susceptible to death by desiccation, are easily washed away or battered by wave action, and otherwise require protection during frequent molts (every 2–4 days).

Five- to eight-month-old crabs were more active than the younger ones, often moving around the perimeter but stopping briefly for several seconds as they passed the screen edge (Herrnkind, 1968*b*). The reduced affinity for objects was also apparent in the field, except when the crabs were desiccated or pursued, although they oriented spontaneously toward landmarks in the absence of other guidance cues.

c. Celestial Cues. Directional responses to celestial cues varied according to the age of the test crabs, with landward preferences becoming more prominent in older specimens. Both wild and laboratory-reared crabs less than 2 months old demonstrated no single persistent directional preference (Fig. 8*a*,*b*). Pronounced, but variable, directional tendencies occurred among slightly older, wild individuals, but again no distinct bearing appeared in the combined data (Fig. 8c). Generally, the broad scatter reflected both large differences in preferred bearings of individuals (Fig. 8d) and more often the absence of any directional preference in others. These latter individuals moved about the chamber in the same manner as apparently unoriented control crabs. Furthermore, the young crabs showed little or no tendency to group as do adults. Hence, the high degree of variability in preferred direc-

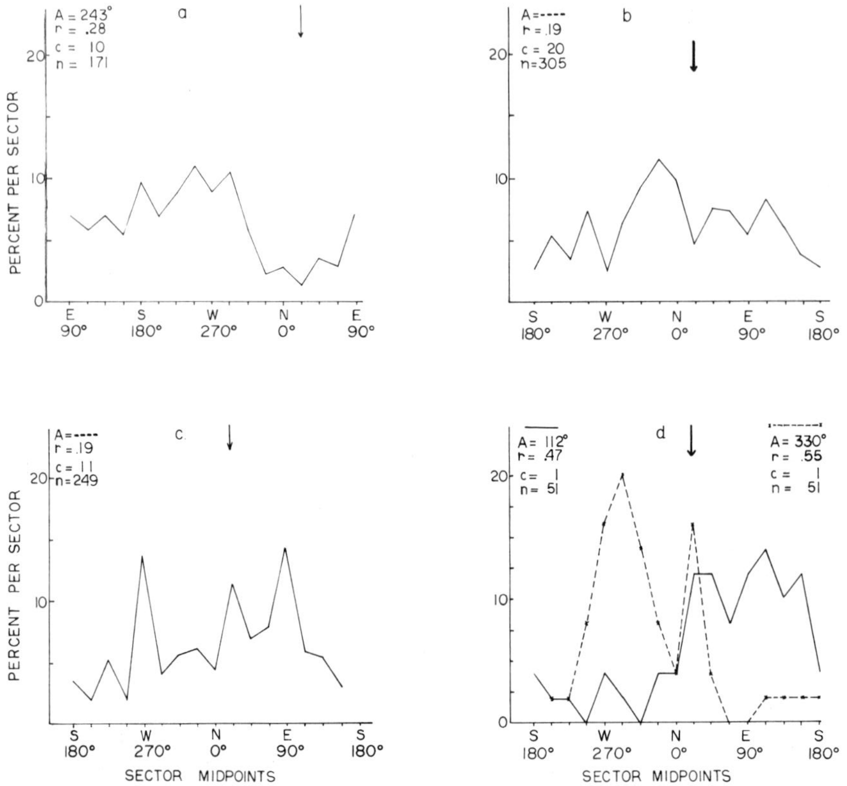

Fig. 8. Distributions of immature sand fiddlers in circular chambers permitting only a view of the clear sky. **(a)** Wild specimens under 2 months old. **(b)** Laboratory-reared specimens under 2 months old. **(c)** Wild crabs 2–5 months old. **(d)** Individual directional preferences of two laboratory-reared crabs under 2 months old. Hypothetical landward (↓) is NNE or 22.5° in each case.

tions by some laboratory-reared crabs, the absence of consistent orientation in others, and the equivalent performance by wild crabs of the same age indicate that young crabs do not spontaneously orient in the landward direction. Additionally, the inconsistency of orientation by young wild crabs implies that celestial cues play a relatively unimportant role in guiding specific locomotory activities of early instars in nature.

Despite the absence of strong directional responses to celestial cues, the distributions are not uniform, showing that the presence of sun and/or sky influences the orientation of young crabs. The stimulus responsible appears to be differential light intensities, resulting from sun angle and reflections off parts of the test chamber. For example, Fig. 9 shows the definite photopositive response of a group of crabs, at this age, in an unevenly illuminated

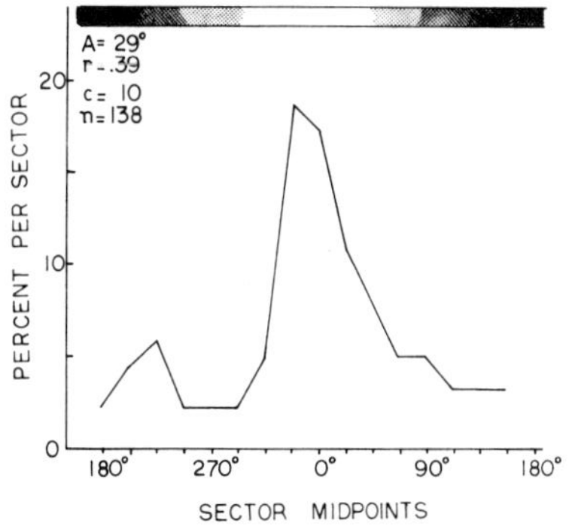

Fig. 9. Photopositive orientation shown by crabs, about 2 months old, immersed in seawater to their carapaces. The test chamber is a 30 cm diameter chamber, the translucent white wall of which is lighted on one side by a microscope illuminator (strip above graph). Scattered distributions, such as those in Fig. 6, occur if the light, but not the infrared (heat) from the illuminator, is blocked by an opaque screen.

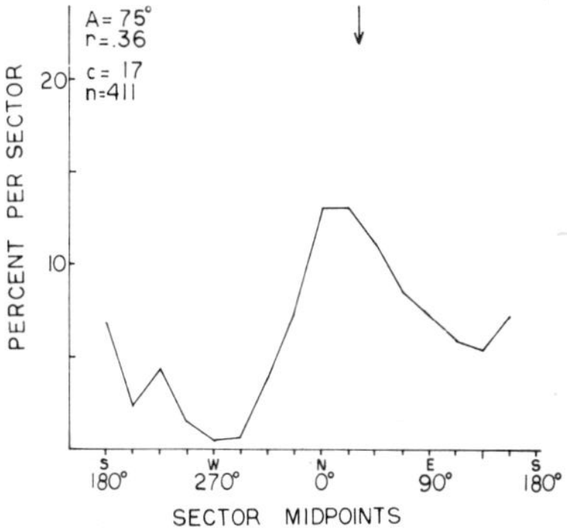

Fig. 10. Distributions shown by wild sand fiddlers 5–8 months old in response to celestial cues. Landward is indicated by the arrow (↓).

chamber. Such positive photo-orientation may induce early instars to move progressively upshore into the comparatively brighter sandy regions of the beach. This, however, is speculative, since the phototactic and photokinetic responses of young crabs were not extensively investigated.

Older crabs, 5–8 months old, oriented generally landward, although the distribution of directional choices was significantly more dispersed (Fig. 10; mean vector = 0.36) than for adults (Fig. 4a; mean vector = 0.93). However, the subadults tended to group together in the chamber and to climb at the wall as do adults. Consequently, their behavior is apparently a manifestation of the celestially oriented landward response, exhibited most clearly by crabs 10 months and older, and distinctly different from the inconsistent responses of crabs younger than 3–4 months. The appearance of strong landward orientation to celestial cues at 9–10 months corresponds closely with the onset of adult social behavior, including maintenance of burrows on the upper beach and feeding in wide-ranging droves. Such activities necessitate, or are aided by, celestial orientation (Altevogt and von Hagen, 1964; Herrnkind, 1968*b*).

d. Celestial Cues vs. Landmarks. The successive ontogenetic changes in response to landmark and celestial cues were also apparent in the screen–sky choice experiments. Wild crabs less than 2 months old approached and aggregated at the screen independent of its position relative to the albeit weak direction preferred in celestial tests (Fig. 11a). Specimens 2–5 months old also approached the screen but did not remain aggregated as strongly as the younger group (Fig. 11b). Crabs 5–8 months old showed variable preferences for both the screen and landmark direction (Fig. 11c). For example, one

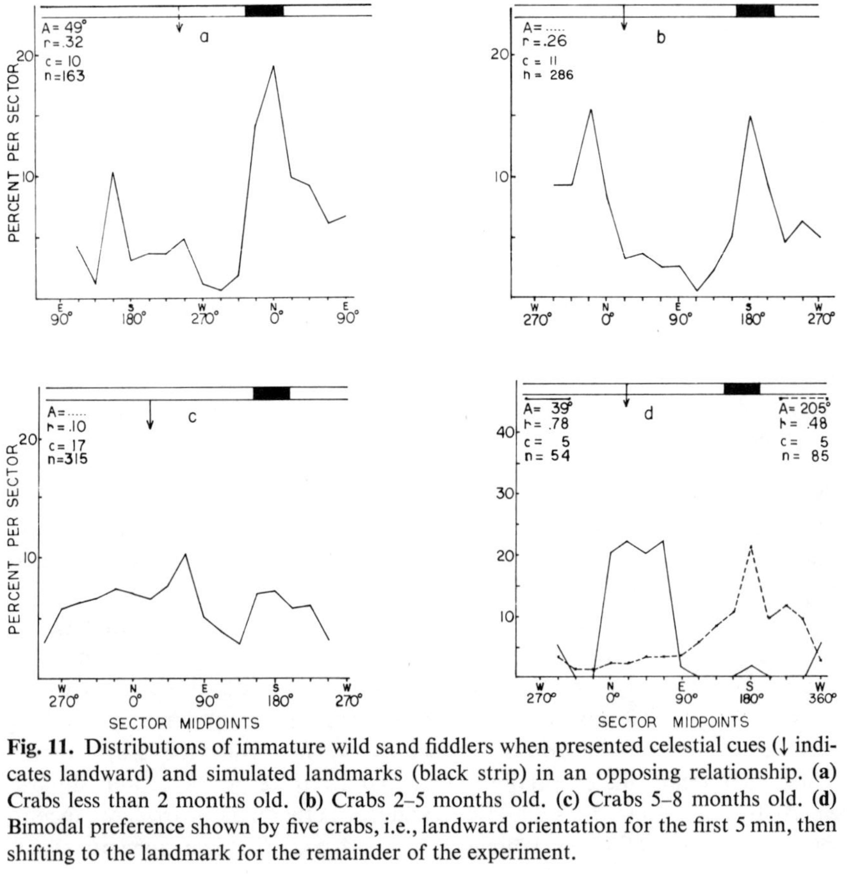

Fig. 11. Distributions of immature wild sand fiddlers when presented celestial cues (↓ indicates landward) and simulated landmarks (black strip) in an opposing relationship. **(a)** Crabs less than 2 months old. **(b)** Crabs 2–5 months old. **(c)** Crabs 5–8 months old. **(d)** Bimodal preference shown by five crabs, i.e., landward orientation for the first 5 min, then shifting to the landmark for the remainder of the experiment.

group oriented landward for 5 min, then moved to the screen, 180° away, for the remainder of the test (Fig. 11d). Most adults in the same test situation orient strongly landward, although this preference can be modified or reversed by desiccation (Herrnkind, 1968*b*).

4. Discussion

The orientational behavior of *U. pugilator* undergoes a progressive transition between the first crab stage and the approximate time of sexual maturity. Early instars do not spontaneously orient by celestial cues in the landward bearing of the parent's habitat and do not show significant effects from experience prior to 2 months of age. Celestial orientation is inconsistent, and probably insignificant, to crabs from this habitat before 3–4 months of age. Rather, the predominant visual orientation during this time is a spontaneous telotaxis toward areas of optical contrast. Celestially oriented escape responses appear by the subadult stage at 5–8 months. A distinct, landward-directed celestial orientation appears at about 1 year of age under conditions of stress in the environment. The telotaxis toward areas of contrast, so predominant in the preadult stages, appears in the adult only under specific conditions. The simple taxes of early instars are eventually dominated by celestial menotaxes in spatial reference to the specific habitat.

Clearly, the appearance of landward orientation by celestial cues during ontogeny is accompanied by a dominating influence of that mode over telotaxis and simple phototaxis. These latter mechanisms are apparently sufficient to guide the short, diffuse movement of the early instars, which probably have no behavioral need of a permanent directional reference. On the other hand, the extensive and complex locomotory activities of adults cannot be effectively guided by simple taxes, and hence they rely upon celestial orientation.

5. Establishment of Directional Preferences

How directional preferences are established is answered, in part, by results of a preliminary study on orientation induced by raising crabs in simulated habitats. The purpose of these experiments was twofold: (1) to determine the orienting effect of polarized light on young crabs and (2) to provide information regarding ecological factors relevant to establishment of directional orientation.

a. Methods. One to five laboratory-reared crabs, less than 1 week old, were placed in simulated habitats consisting of a sand shore and seawater in a 30 cm diameter plastic pan (Herrnkind, 1968*a*,*b*). An overhead polarizer (Polaroid HN38), a diffuser, and a 25 or 60 W light source were situated above the pan. The crabs were left undisturbed except for occasional observation or feeding. Light was held constant, and the plane of polarization was set

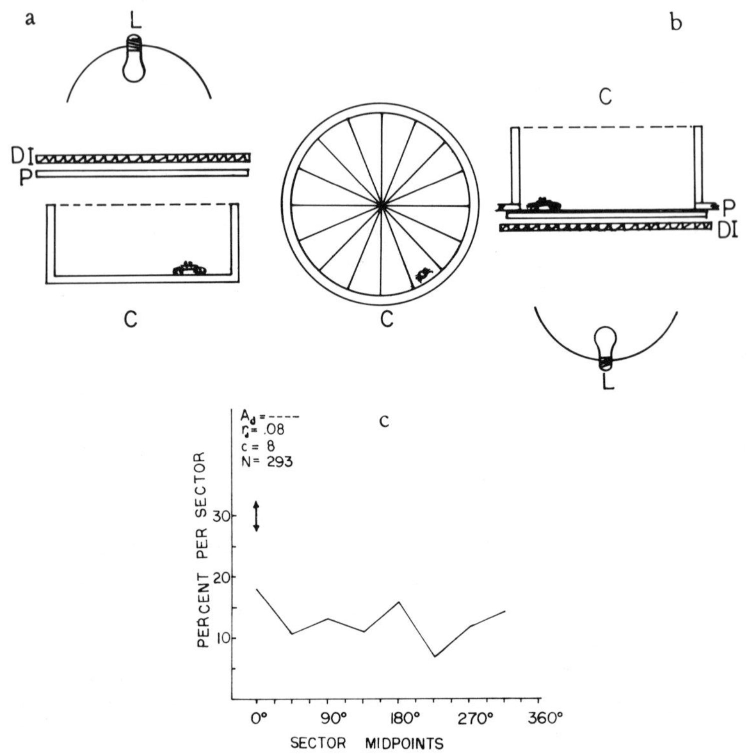

Fig. 12. **(a)** Left: Arrangement of 30 cm chamber (C), overhead 60 W light (L), diffuser (DI), and polarizer (P) for testing responses to polarized light. Center: View from above or below showing transparent floor with gridlines drawn every 22.5°. **(b)** Right: Arrangement of components for testing responses to intensity artifacts caused by the differential reflection of polarized light from the chamber walls. Equivalent intensity artifacts are produced on the pan sides in either arrangement, but crabs do not visually perceive optical phenomena directly below them. The crabs are placed directly on the polaroid in the arrangement on the right. **(c)** Distribution exhibited by crabs in the latter arrangement. No directional preference is indicated. The symbol (⟷) represents the plane of polarization parallel to the 0–180° axis.

either parallel or perpendicular to the "shoreline." Pan walls were either sprayed a uniform flat white or surrounded with wrinkled black vinyl film to reduce intensity artifacts. Tests were conducted in the same apparatus or a duplicate in which a small amount of seawater (several millimeters depth) was provided to prevent desiccation (Fig. 12a).

Crabs moved mostly around the perimeter of the test pan, or across the pan floor, stopping briefly about every 10–20 sec. Their position was recorded whenever they stopped, each test period lasting for 2–5 min. Thereafter, the polaroid was turned through a significant angle. The angle of the polarization plane was determined with a calibrated polaroid sheet each recording period.

It was postulated that if the crabs associated the shore-water axis with the plane of polarization in the pan, they would, like the adults, orient in the same relationship to polarized light in a pan without a shore. Crabs held either under diffuse depolarized light or in pans without shores were tested by the procedures described to determine whether any polarotactic orientation developed in the absence of polarization stimuli or shore–water reference. To be sure that intensity artifacts, resulting from differential reflection of polarized light off the chamber walls, did not influence orientation, the same test procedure was performed with the polarized light coming from below (Fig. 12b). Hence, the intensity artifacts were equivalent, but the polarized light did not enter the crabs' eyes directly. Eight crabs, 1–2 months old, showed nearly uniform dispersion under these conditions (Fig. 12c), indicating they did not orient to intensity artifacts.

b. Results. The data did not reveal any general shore–water orientation to polarized light in crabs less than 2 months old (Fig. 13a). The dispersion shown, as well as the behavior of the test crabs, resembled that of the control crabs 2 weeks to 5 months old (Fig. 12c and 13b). Definite orientation, perpendicular to the shoreline relative to the polarization plane, appeared in five crabs 2–3 months old, as shown by their cumulative distribution (Fig. 13c). This preferential response occurred in crabs 2–3 months younger than those wild crabs exhibiting strong landward orientation on a real beach.

c. Discussion. It is evident from these results that individuals between the first and tenth instar do not innately orient in any single preferential direction relative to plane polarized light, Consequently, the directional tendencies exhibited by crabs at this age under the clear sky probably do not reflect polarotaxis but rather are the result of photo-orientation to light intensity patterns, as suggested previously.

Results from the simulated habitat experiments provide the basis for hypothesizing the processes responsible for transition of orientational modes during ontogeny and for the establishment of directional preferences. Perhaps the clearest way to relate the processes to be discussed in the following is to consider them in light of the question of why young crabs do not show landward-directed, celestial menotaxes as do adults.

(1) The behavioral repertoire of early instars, up to several weeks of age, does not include directional locomotory patterns which can be associated with celestial cues; hence, no preferred direction is established. This is supported by behavioral observations on laboratory-reared crabs and by the observations on wild crabs in the field by myself and others (Hyman, 1920, 1922). First stage, laboratory-reared crabs in simulated habitats do not migrate onto the sand immediately, but remain underwater much of the time. Individuals

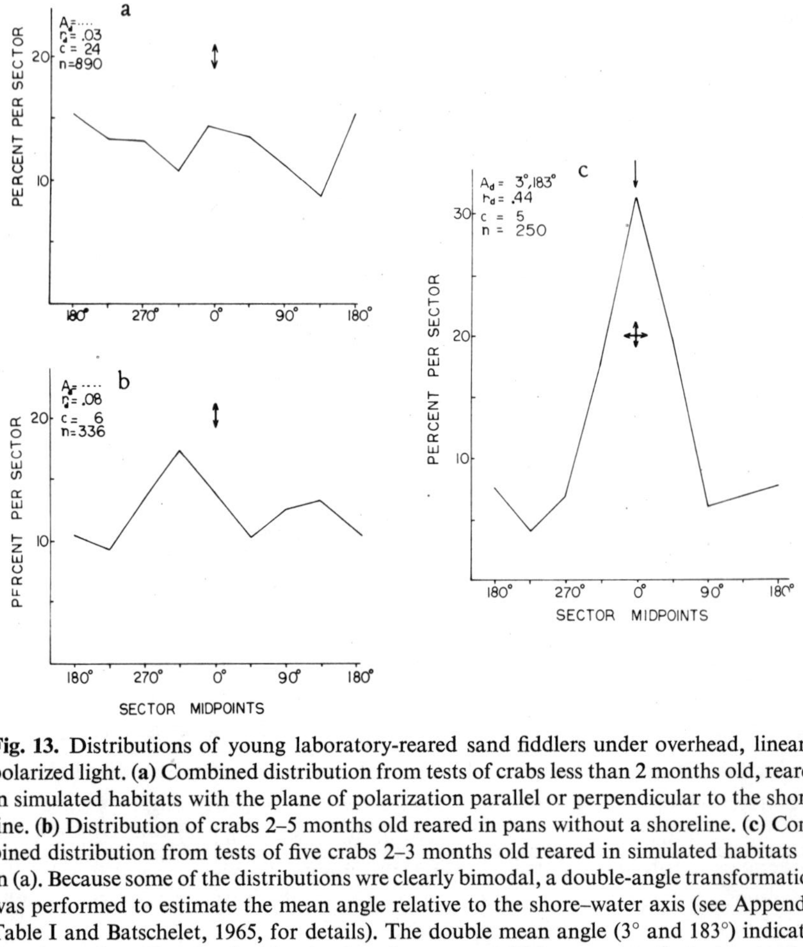

Fig. 13. Distributions of young laboratory-reared sand fiddlers under overhead, linearly polarized light. **(a)** Combined distribution from tests of crabs less than 2 months old, reared in simulated habitats with the plane of polarization parallel or perpendicular to the shoreline. **(b)** Distribution of crabs 2–5 months old reared in pans without a shoreline. **(c)** Combined distribution from tests of five crabs 2–3 months old reared in simulated habitats as in (a). Because some of the distributions wre clearly bimodal, a double-angle transformation was performed to estimate the mean angle relative to the shore–water axis (see Appendix Table I and Batschelet, 1965, for details). The double mean angle (3° and 183°) indicates that the preferred bearing is either landward or seaward (↓) in respect to the incident polarization plane. The symbol (⟷) indicates that crabs were exposed to incident polarization either parallel or perpendicular to the shore–water axis.

between the third and fifth instar begin to emerge onto the shore for longer periods and to dig burrows. Fifth to tenth stage crabs dig and maintain burrows with a daily periodicity but often feed in the water or flee there if disturbed. The field behavior is somewhat analogous; the first three instars, at least, cling to *Spartina* stalks, wander diffusely about the wet portions of the lower beach, or remain submerged in the shallow tide pools. These crabs undergo no apparent locomotory patterns spatially relatable to the plane of polarization relative to the shoreline until after the first 2–3 weeks.

(2) After directional locomotory patterns appear at about 3 weeks to 1

month, the establishment of a preferred bearing relative to celestial guidance cues depends, as in adults, upon the clarity and stability of a shore–water configuration. Young, laboratory-reared crabs traverse a shoreline in permanent spatial relationship to the polarization plane. In addition, the plane of polarization is constant and the physical environment is nearly invariant, particularly compared to the natural intertidal sand beach. On the other hand, wild crabs in the latter habitat experience changing conditions with each tide and a variety of microenvironmental shore–water directions, even on a "straight" shoreline, since they move about within relatively small regions rather than over the whole beach as do adults. Hence, the laboratory-reared, 2- to 3-month-old crabs are able to establish a single celestial reference appropriate to their habitat, whereas wild crabs of the same age and physical development, as a group, cannot do so until the fifth or sixth month.

(3) Locomotory ability and resistance to aerial exposure are inextricably related to the effect of shoreline characteristics on directional preference. The younger the crab, the slower its walking speed and the greater its susceptibility to desiccation; it simply may not be able to cover the area required to "recognize" the general landward bearing. Those individuals which do establish preferred directions may differ in their compass bearing, depending on the characteristics of their beach. This is suggested by results of celestial tests on wild specimens where some individuals 3–4 months old demonstrated strong but divergent headings. By adulthood, crabs not only can move faster and withstand increased exposure, but they also exhibit specific behavioral activities (e.g., drove feeding) in which they traverse the full extent of the beach.

(4) Developmental factors, independent of natural environmental variations and experience, may actually set limits on orientational capabilities throughout the immature period. For example, the visual and supporting neurosensory systems may require maturation before young crabs can perceive polarized light (Korte, 1966) or accurately fixate and detect changes in sun position (Horridge, 1966*a*). Korte's (1966) study of vision in *U. tangeri* shows that acuity decreases markedly in successively younger crabs. This correlates with the lesser number of ommatidia, and consequent increase in visual angle subtended by each, in the younger specimens. Also, any summation process might require an increase in the number of polarization-sensitive, apical ommatidia before perception of the polarized light is possible. The ability to establish celestial orientation may also be dependent upon maintaining a stable visual field, usually assumed to be a function of eyestalk compensatory adjustment (Cohen and Dijkgraaf, 1961). Until 3 or 4 months old, early instars have relatively fixed eyestalks (Hyman, 1922) and may be unable to fixate sun position or the zenithal polarization while moving over uneven substrate. Any or all of the above sensory inadequacies could have

resulted in the observed lack of directional orientation to polarized light in laboratory-reared crabs between 2 weeks and 3 months of age. Further research on development of sensory processes and related behavior during ontogeny will hopefully clarify this issue.

(5) The functional capability of the orientation clock may also require maturation before young crabs can compensate for azimuthal changes in celestial cues. The adoption of orientation to polarized light in simulated habitats may not reflect a similar capability of young in nature, since the angle of polarization was held constant rather than rotated 15°/hr. Thus, it is unknown whether the young crabs can effectively compensate for this change. Rhythmic processes during ontogeny in *Uca* remain to be studied in detail, although initial observations suggest a daily or circadian periodicity in burrow digging by laboratory-reared crabs held in constant conditions (Herrnkind, 1968*c*).

In overview, the present study establishes that maturational and experiential factors operate in adjusting orientation in *U. pugilator* during ontogeny. The resulting oriented behavior depends upon the interaction of the crabs' organismic capabilities, at each developmental level, with the environmental stresses imposed on those stages in a particular habitat. Therefore, as development proceeds, the biological components change, adapting each successive stage for a different ecological situation, while at the same time providing the organismic capacity to modify orientation in accord with those new environmental characteristics.

IV. GENERAL DISCUSSION

Studies on numerous shore-living arthropods, as summarized in Table III, demonstrate the ubiquity and adaptive significance of orientational mechanisms by which these forms maintain themselves within suitable life zones, accomplish feeding migrations, and direct escape responses. The typical bearing during these activities is perpendicular to the run of the shore, i.e., landward or seaward depending on the species and triggering stimuli. Major exceptions to this are the burrow dwellers such as ghost crabs and fiddler crabs, which sometimes preferentially return to their den irrespective of its compass bearing. In any event, landward–seaward locomotory movements are necessities for existence in the beach environment, since habitable regions and sanctuary for each species occur only in specific zones, while sources of food are often deposited higher or lower on the shore some distance away.

The most common mechanism by which this orientation can be accomplished is the celestial compass, a time-compensating menotaxis guided by

Table III. Summary of Orientational Characteristics of Selected Shore-Living Arthropods Discussed in Text

		Orientation mechanisms				
		Visual		Nonvisual		Establishment of directional preferences guided by celestial cues
Animal group	Oriented activities	Sun compass	Lunar compass	Landmarks		
Amphipods						
Tailtrus saltator and other species with similar behavior–Talorchestia deshayesei, T. longicornis, T. megalophthalma, Orchestia mediterranea, Orchestoidea benedicti	Nocturnal migrations landward or seaward; escape responses by day—landward when on wet substrate, seaward when on dry substrate.	Time-compensating menotaxis to sun or polarized sky light.	Time-compensating menotaxis on lunar cycle.	Positive telotaxis to areas or objects of optical contrast.	Anemotaxis in absence of visual cues; indication of landward–seaward orientation to unknown cue(s) in total darkness.	"Innate," maturational landward–seaward preference; learning appears to be an adjusting mechanism.
Orchestoidea corniculata	As above.	Yes.	Yes; hypothetically, lunar orientation requires resetting "clock" each night.			
Isopods						
Tylos latreilli	Landward migrations nocturnally; escape responses as above.	Yes; as above.	Yes.			
Tylos punctatus	Seaward migrations nocturnally.	No; negative phototaxis only.	No.		Orientation to maximum slope inclination: upslope if wet, downslope if dry.	

Table III (Cont'd)

Animal group	Oriented activities	Orientation mechanisms: Visual: Sun compass	Orientation mechanisms: Visual: Lunar compass	Orientation mechanisms: Nonvisual: Landmarks	Establishment of directional preferences guided by celestial cues
Wolf spiders					
Arctosa variana	Escape response to land when immersed.	Yes; as above.			Spontaneous northerly orientation in young; landward orientation learned in few days if provided sloped shoreline; new escape directions can be learned thereafter.
Decapods					
Uca pugilator *U. tangeri*	Seaward migration diurnally at low tide; landward escape responses when disturbed (seaward sometimes); landward orientation when immersed; rapid return to burrow from up to 1 m.	Yes; as above.		Yes; as above, in young individuals especially; adults can learn to orient to objects.	No spontaneous, "innate" directional preference for landward; directional preferences learned after several months of age in relation to physiography of habitat; kinesthetic reorientation to burrow independent of visual cues.

Ocypode ceratophthalma	Movements traversing lower beach to feed.	Yes; as above in respect to polarized light.		Yes; may locate burrow by placement of nearby objects.	
	Escape responses landward, seaward, or to burrow.				Yes; kinesthesis probably similar to *Uca*.
O. saratan				Orientation to sand cones built by males.	
Goniopsis cruentata	Movements around mangrove roots and perpendicular to shore.	Yes. As above.			

the sun and/or polarized sky light (and in some species the moon as well). Nearly all species tested demonstrate this capability, excepting the isopod *Tylos punctatus* (Hamner *et al.*, 1968), which has an effective substitute mechanism for its particular habitat. The most obvious difference in operation of the compass in each species, under natural conditions, lies in the triggering stimuli inducing the oriented movements.

The ubiquity of the compass may be attributable to the selective advantage it provides. It enables an animal to establish and maintain the appropriate course when out of sensory contact with the "goal." Shore crustaceans probably do not perceive habitable zones, unless in direct physical contact with them, and even such a highly visual form as *Uca* apparently does not see its burrow, even when within 1 m distance. Added to this are movements through grass, debris, and water which also obscure spatial references and further emphasize the utility of the celestial navigating mechanism. Secondly, celestial cues provide the most predictable and accurate navigational information available to the animals. For example, (1) the sun and sky polarization provide a stable beacon for daytime movements, and these cues are permanent and not subject to displacement or modification by erosion; (2) the northerly or southerly bias of the sun's path in either hemisphere provides a compass reference; (3) the sun's movements provide a direct measure of influential temporal phenomena such as photoperiod, time of day, season, etc.; (4) the sun acts both as a guidepost and directly, or indirectly, as a Zeitgeber for various rhythmic components of the compass mechanism and oriented activities. Finally, celestial cues, particularly by day, are available under most environmental conditions. The sun is obscured only by dense overcast, by fog, by storms, and briefly by passing clouds. In this last situation, available polarized light, causally related to sun position, provides essentially the same navigational information. These cues are also perceptible to animals moving through grass and underwater. Hence, animals possessing a celestial compass are seldom without spatial reference (by day).

As important as the mechanism itself are the processes by which a preferred bearing is established in different species. The shore-living forms, with similar compass orientation, show clear differences related to their mode of development, life history and ecology. The talitrid amphipods present a situation unusual in arthropods, in which instars, upon leaving the marsupium, demonstrate a spontaneous tendency to orient in the compass reference of the parent. Naive, laboratory-reared lycosid spiders exhibit a northerly compass orientation soon after birth, but this is presumably a manifestation of the endogenous mechanism without an adaptive reference. Wolf spiders and fiddler crabs develop orientational preferences after experience in the environment. A strong, genetically fixed, directional tendency could be maladaptive in these latter forms, since the parent may change preferred bearings during its lifetime or the offspring may develop in completely different areas. In all these groups, however, the young orient differently than older, experienced individuals and adults. All show relatively strong photo- and telotactic responses which, in *Arctosa* and *Uca*, dominate other types of orientation until later stages. The adaptive function of these responses is in question, although the telotaxis facilitates shelter seeking in young fiddler crabs at stages highly susceptible to environmental stress. The orientational repertoire demonstrated during ontogeny in each group reflects the different modes of life. Thus, wolf spiders and amphipods show strong, landward escape responses throughout most of their life, while fiddler crabs do not. Both the orientation mechanisms and types of oriented activities in *Uca* change markedly between early instars and the adult stages. Nevertheless, compass orientation in wolf spiders and fiddler crabs is shaped in large part by the physiography of their habitats. Considerable plasticity of the orientational repertoire is indicated by the variety and modifiability of directional preferences in both animals. The learning processes seem to require locomotion relative to some consistent physical reference which is latently associated with celestial guideposts.

Another common visual mechanism involves orientation by landmarks. Most animals in which this is observed, namely *Talitrus*, *Uca*, and *Arctosa*, exhibit telotactic-like orientation directly toward a rather wide range of objects or shapes. Only *Ocypode* and *Uca* thus far appear to actually discriminate and orient to specific forms, although wolf spiders almost certainly have this visual capability. Generally, landmarks provide an adequate guidepost in the absence of celestial cues and may serve as sources of refuge or habitable conditions; e.g., the shore is both guidance cue and goal for wolf spiders, while orientation toward landmark vegetation often brings fleeing fiddler crabs through their burrow area.

Nonvisual mechanisms serve to augment, as well as to replace, visual orientation under certain conditions. The commonest nonvisual orientation occurs in response to gravitational cues on slopes, although these usually serve as a postural mechanism and are known to guide directional locomotion only in *Arctosa* and *Tylos punctatus*. The usefulness of this mode is restricted to regions of relatively constant beach slope, since slight irregularities easily confound directional information. Another mechanism, found in *Uca*, *Ocypode*, and probably other burrow dwellers, is kinesthetic orientation by proprioceptive cues, perhaps in conjunction with tactile cues. This mechanism is especially functional to burrow dwellers in enabling them to accurately return to a burrow although it is not visible. Kinesthesis may further serve in most arthopods to correct for deviations in course caused by detours (Barnwell, 1965; Schöne, 1965; von Hagen, 1967). Recent evidence of guidance in amphipods by as yet unpostulated mechanisms requires further elucidation to demonstrate any functional significance. A nonvisual mechanism, equivalent to the celestial compass, would be especially valuable to these nocturnally active species.

Thus far, the discussion has concerned the responses of animals to one class of guidance stimuli when others are excluded. For the most part, this is how the research has been conducted, since the primary goal has often been to analyze a single mechanism. Yet, as Edney (1968) points out, "animals in nature are subjected to a multiplicity of physical and physiological variables, and their behavior at any one time is the result of a complex interplay of all these factors." In other words, one cannot say with certainty that a fiddler crab which *can* orient strongly landward by a sun compass when in a level, featureless, circular chamber excluding all landmarks actually *does* so when fleeing landward on the beach while moving upslope, toward prominent landmarks, in the company of several conspecifics. In fact, when landmark and celestial cues are presented in some opposing relationship, either in the field or laboratory, crabs demonstrate selectivity toward one or the other. Thus, a sun compass capability in *Uca* does not provide adequate evidence that it is the only, or even the primary, mechanism operating in the field. Orientation by other species (mentioned herein) is probably more complex, variable, and labile than the available literature might suggest.

A clear understanding of the operation and function of directional orientation is best demonstrated in relation to the environmental conditions faced by the animal (Edney, 1968), since the orientation mechanisms are both adapted to those stimulus conditions and controlled by them. From a systems analysis standpoint (a useful way to approach this problem), there are mutiple input sources each with information to be integrated into but a single output (Waterman, 1966). Thus experiments should test responses over a known range of single cues and, also, with multiple cues in opposition. Present availability of constant-recording monitors, as well as the capacity to computer-analyze large quantities of multivariate data, should sophisticate and facilitate this approach (Wenner *et al.*, 1967). This approach must also integrate both field experimental techniques and experiments under more controlled conditions; interpretation of results from tests in artificial conditions should eventually be predictive of behavior in the field (successful examples of this approach are found in Hasler, 1967; Jander, 1957; von Frisch, 1967).

APPENDIX

The parameters of the distributions made by test crabs were determined according to methods described by Batschelet (1965) and discussed also by Papi and Tongiorgi (1963). The mean direction (A) and the mean vector (r), representing the strength of grouping about the mean direction, as well as the number of test animals (c) and the total number of positions recorded (n), are given in each of the text figures and in Appendix Table I. In addition, the table provides the number of positions (No.) and proportion (%) for each of the 16 sectors, the proportion (%) providing the values graphed in each of the text figures.

Statistical estimates of the presence and degree of directional preference exhibited by each distribution are provided by chi-square values for each sector, as well as for the total of all 16 sectors, and by the value of the Rayleigh test statistic, z. In the chi-square, probabilities are given for each sector in which the observed number of positions recorded exceeds the expected value, as well as the probability based on the total distribution. The Rayleigh test values given are based on values derived both from the distribution of all positions recorded (n) and from the distribution as exhibited by the number of crabs tested (c). In each case, a z value less than z_p, 0.5 does not reject the hypothesis of a uniform distribution. The data for Figs. 12 and 13 represent a special case, since the Rayleigh test and the method for determining mean angle (A) are applicable only to unimodal distributions. Bimodal groupings, with peaks 180° apart, were often recorded during polarized-light experiments necessitating a "double-angle" transformation of the distribution to a single mode (see Batschelet, 1965, for details). The doubling of each angle from any reference point results in a unimodal distribution. The new mean vector and mean angle are then treated statistically. The values of A_d, r_d, and z for bimodal distributions were computed from "double-angle" values.

Appendix Table I. Parameters of Test Crab Distibution (Fig. 3a)

	Fig. 3a (left) $c = 5$ $0° =$ north				Fig. 3a (right) $c = 5$ $0° =$ north			
Sector	No.	%	χ^2	$(>\chi^2$; 1 df)	No.	%	χ^2	$(>\chi^2$; 1 df)
0°	37	37	151.29	($\ll$0.001)	12	12	5.29	(0.02)
22.5°	9	9	1.21		11	11	3.61	(0.10)
45°	0	0	6.25		4	4	0.81	
67.5°	0	0	6.25		6	6	0.01	
90°	0	0	6.25		1	1	4.41	
112.5°	0	0	6.25		2	2	2.89	
135°	0	0	6.25		1	1	4.41	
157.5°	0	0	6.25		1	1	4.41	
180°	0	0	6.25		5	5	0.25	
202.5°	0	0	6.25		0	0	6.25	
225°	0	0	6.25		3	3	1.69	
247.5°	0	0	6.25		1	1	4.41	
270°	1	1	4.41		15	15	12.25	(0.001)
292.5°	4	4	0.81		13	13	7.29	(0.01)
315°	17	17	18.49	($<$0.001)	16	16	15.21	(0.001)
337.5°	32	32	106.09	($\ll$0.001)	9	9	1.21	
	$n = 100$		$\chi^2_T = 344.80$	($\ll$0.001; 15 df)	$n = 100$		$\chi^2_T = 74.40$	($<$0.001; 15 df)
	$r = 0.93$	$z_c = 4.32$ $(>z_p, 0.01)$			$r = 0.51$	$z_c = 1.31$ $(<z_p, 0.05)$		
	$A = 354°$	$z_n = 86.49$ $(>z_p, 0.0001)$			$A = 348°$	$z_n = 26.01$ $(>z_p, 0.0001)$		

Appendix Table I. Cont'd (Fig 4a–c)

Fig. 4a $c = 30$ $0° =$ North				Fig. 4b $c = 30$ $0° =$ North				Fig. 4c $c = 10$ $0° =$ north			
No.	%	χ^2	$(>\chi^2; 1\ df)$	No.	%	χ^2	$(>\chi^2; 1\ df)$	No.	%	χ^2	$(>\chi^2; 1\ df)$
35	22	61.80	(<0.001)	2	1	9.00		3	7	0.05	
62	39	268.07	(≪0.001)	2	1	9.00		4	10	0.72	
44	27	114.46	(≪0.001)	10	5	0.57		5	12	2.15	
12	8	0.37		8	4	1.73		1	2	1.01	
0	0	10.06		11	5	0.22		3	7	0.05	
2	1	6.46		14	7	0.13		2	5	0.15	
0	0	10.06		18	9	2.22		2	5	0.15	
0	0	10.06		25	12	11.95	(0.001)	4	10	0.72	
0	0	10.06		35	17	39.24	(<0.001)	4	10	0.72	
0	0	10.06		24	12	10.09	(0.001)	3	7	0.05	
0	0	10.06		24	12	10.09	(0.001)	3	7	0.05	
0	0	10.06		9	4	1.07		2	5	0.15	
0	0	10.06		10	5	0.57		2	5	0.15	
0	0	10.06		5	3	4.66		0	0	2.63	
0	0	10.06		3	2	7.40		1	2	1.01	
6	4	1.64		3	2	7.40		3	7	0.05	
$n = 161$ $\chi_T^2 = 553.43$			(≪0.001; 15 df)	$n = 203$ $\chi_T^2 = 115.34$			(≪0.001; 15 df)	$n = 42$ $\chi_T^2 = 9.81$			. . .
$r = 0.93$ $z_c = 25.95$				$r = 0.49$ $z_c = 7.20$				$r = 0.12$ $z_c = 0.14$			
$(>z_p, 0.0001)$				$(>z_p, 0.001)$				$(<z_p, 0.05)$			
$A = 49°$ $z_n = 139.25$				$A = 197°$ $z_n = 48.74$				$A = . . .$ $z_n = 0.60$			
$(>z_p, 0.0001)$				$(>z_p, 0.0001)$				$(<z_p, 0.05)$			

Appendix Table I. Cont'd (Fig. 6a–c)

Fig. 6a $c = 3$				Fig. 6b $c = 5$				Fig. 6c $c = 5$			
No.	%	χ^2	($>\chi^2$; 1 df)	No.	%	χ^2	($>\chi^2$; 1 df)	No.	%	χ^2	($>\chi^2$; 1 df)
11	7	0.20		5	6	0.02		6	7	0.34	
9	6	0.04		8	9	1.36		7	8	0.37	
11	7	0.20		7	8	0.54		7	8	0.37	
7	5	0.72		4	5	0.32		1	1	3.74	
6	4	1.37		4	5	0.32		8	9	1.07	
9	6	0.04		4	5	0.32		5	6	0.06	
11	7	0.20		2	2	2.07		6	7	0.03	
13	8	1.18		6	7	0.09		4	5	0.44	
10	7	0.01		8	9	1.36		5	6	0.06	
11	7	0.20		6	7	0.09		10	11	3.54	(0.10)
5	3	2.22		4	5	0.32		4	5	0.44	
8	5	0.27		9	9	2.60		7	8	0.37	
7	5	0.72		4	5	0.32		3	3	1.18	
9	6	0.04		2	2	2.07		4	5	0.44	
14	9	1.99		4	5	0.32		10	11	3.54	(0.10)
13	8	1.18		8	9	1.36		2	2	2.28	
$n = 154$		$\chi^2_T = 10.57$	...	$n = 85$		$\chi^2_T = 13.45$	...	$n = 89$		$\chi^2_T = 17.97$	...
$r = 0.03$ $z_c = 0.003$ ($<z_p$, 0.05)				$r = 0.03$ $z_c = 0.003$ ($<z_p$, 0.05)				$r = 0.02$ $z_c = 0.002$ ($<z_p$, 0.05)			
$A = \ldots$ $z_n = 0.14$ ($<z_p$, 0.05)				$A = \ldots$ $z_n = 0.08$ ($<z_p$, 0.05)				$A = \ldots$ $z_n = 0.04$ ($<z_p$, 0.05)			

Appendix Table I. Cont'd (Fig. 7a–c)

Fig. 7a $c = 20$				Fig. 7b $c = 9$				Fig. 7c $c = 10$			
No.	%	χ^2	($>\chi^2$; 1 df)	No.	%	χ^2	($>\chi^2$; 1 df)	No.	%	χ^2	($>\chi^2$; 1 df)
65	25	149.10	($\ll$0.001)	62	29	175.50	($\ll$0.001)	33	18	41.11	(<0.001)
8	3	4.05		14	7	0.02		21	12	8.14	(0.005)
8	3	4.05		20	9	3.20		11	6	0.01	
9	4	3.11		2	1	9.74		15	8	1.16	
9	4	3.11		11	5	0.44		3	2	6.17	
10	4	2.29		8	4	2.20		7	4	1.68	
9	4	3.11		3	1	8.11		13	7	0.23	
13	5	0.58		1	1	11.51		6	3	2.54	
13	5	0.58		3	1	8.11		10	6	0.17	
8	3	4.05		3	1	8.11		5	3	3.57	
8	3	4.05		1	1	11.51		7	4	1.68	
11	4	1.60		3	1	8.11		4	2	4.78	
12	5	1.03		5	2	5.30		2	1	7.73	
4	2	9.06		2	1	9.74		11	6	0.01	
4	2	9.06		16	7	0.49		14	8	0.61	
66	26	155.25	($\ll$0.001)	61	28	168.35	($\ll$0.001)	20	11	6.54	(0.02)
$n = 257$		$\chi^2_T = 354.05$	($\ll$0.001; 15 df)	$n = 215$		$\chi^2_T = 430.43$	($\ll$0.001; 15 df)	$n = 182$		$\chi^2_T = 86.13$	(<0.001; 15 df)
$r = 0.38$	$z_c = 2.89$	($>z_p$, 0.10)		$r = 0.68$	$z_c = 4.16$	($>z_p$, 0.05)		$r = 0.34$	$z_c = 1.16$	($<z_p$, 0.05)	
$A = 13°$	$z_n = 37.11$	($>z_p$, 0.0001)		$A = 21°$	$z_n = 99.42$	($>z_p$, 0.0001)		$A = 32°$	$z_n = 21.04$	($>z_p$, 0.0001)	

Appendix Table I. Cont'd (Fig. 8a–c)

Fig. 8a $c = 10$ $0° =$ north				Fig. 8b $c = 20$ $0° =$ north				Fig. 8c $c = 11$ $0° =$ north			
No.	%	χ^2	($>\chi^2$; 1 df)	No.	%	χ^2	($>\chi^2$; 1 df)	No.	%	χ^2	($>\chi^2$; 1 df)
5	3	3.03		30	10	6.28	(0.025)	11	4	1.34	
2	1	7.06		14	5	1.34		28	11	9.94	(0.005)
6	3	2.06		23	8	0.81		17	7	0.13	
5	3	3.03		22	7	0.45		19	8	0.76	
12	7	0.16		16	5	0.49		35	14	24.28	(<0.001)
10	6	0.04		25	8	1.85		14	6	0.16	
12	7	0.16		18	6	0.06		13	5	0.42	
9	5	0.27		11	4	3.41		7	3	4.71	
17	10	3.73	(0.10)	8	3	6.42		9	4	2.77	
12	7	0.16		16	5	0.49		5	2	7.17	
15	9	1.74		10	3	4.31		13	5	0.42	
19	11	6.47	(0.025)	22	7	0.45		5	2	7.17	
15	9	1.74		8	3	6.41		34	14	21.84	(<0.001)
18	11	5.00	(0.05)	19	6	0.06		10	4	1.99	
10	6	0.64		28	9	4.19	(0.05)	14	6	0.16	
4	2	4.18		35	11	13.32	(0.001)	15	6	0.02	
$n = 171$		$\chi^2_T = 38.87$	(0.001: 15 df)	$n = 305$		$\chi^2_T = 50.30$	(<0.001; 15 df)	$n = 249$		$\chi^2_T = 83.27$	(<0.001; 15 df)
$r = 0.28$				$r = 0.19$	$z_c = 0.72$ ($<z_p$, 0.05)			$r = 0.19$			
$A = 243°$				$A = \ldots$	$z_n = 11.01$ ($>z_p$, 0.001)			$A = \ldots$			

Appendix Table I. Cont'd (Fig. 8d)

Fig. 8d(left) $c = 1$ 0° = north				Fig. 8d(right) $c = 1$ 0° = north			
No.	%	χ^2	$(>\chi^2; 1\ df)$	No.	%	χ^2	$(>\chi^2; 1\ df)$
2	4	0.44		2	4	0.44	
6	12	2.48		8	16	7.27	(0.01)
6	12	2.48		2	4	0.44	
4	8	0.21		0	0	3.19	
6	12	2.48		0	0	3.19	
7	14	4.56	(0.05)	1	2	1.50	
5	10	1.03		1	2	1.50	
6	12	2.48		1	2	1.50	
2	4	0.44		1	2	1.50	
1	2	1.50		1	2	1.50	
1	2	1.50		1	2	1.50	
0	0	3.19		4	8	0.21	
2	4	0.44		8	16	7.27	(0.01)
1	2	1.50		10	20	14.56	(0.001)
0	0	3.19		7	14	4.56	(0.05)
2	4	0.44		4	8	0.21	
$n = 51$		$\chi^2_T = 28.37$	(0.02; 15 df)	$n = 51$		$\chi^2_T = 50.33$	(<0.001; 15 df)
$r = 0.47$		$z_c = 0.22$	$(<z_p, 0.05)$	$r = 0.55$		$z_c = 0.30$	$(<z_p, 0.05)$
$A = 112°$		$z_n = 11.27$	$(>z_p, 0.0001)$	$A = 330°$		$z_n = 15.43$	$(>z_p, 0.0001)$

Appendix Table I. Cont'd (Fig. 9 and 10)

Fig. 9 $c = 10$				Fig. 10 $c = 17$ 0° = north			
No.	%	χ^2	$(>\chi^2; 1\ df)$	No.	%	χ^2	$(>\chi^2; 1\ df)$
24	17	27.41	(<0.001)	56	14	35.77	(<0.001)
15	11	4.72	(0.05)	56	14	35.77	(<0.001)
11	8	0.65		46	11	16.06	(0.001)
7	5	0.31		35	9	3.38	(0.10)
7	5	0.31		30	7	0.72	
5	3	1.52		24	6	0.11	
5	3	1.52		22	5	0.53	
5	3	1.52		30	7	0.72	
3	2	3.67		28	7	0.21	
6	4	0.80		9	2	10.84	
8	6	0.05		18	4	2.30	
3	2	3.67		6	2	15.09	
3	2	3.67		2	1	21.84	
3	2	3.67		3	1	20.04	
7	5	0.31		16	4	3.6	
26	19	35.00	(<0.001)	30	7	0.72	
$n = 138$		$\chi^2_T = 88.78$	(<0.001; 15 df)	$n = 411$		$\chi^2_T = 167.76$	(≪0.001; 15 df)
$r = 0.39$		$z_c = 1.52$	$(<z_p, 0.05)$	$r = 0.36$		$z_c = 2.20$	$(<z_p, 0.05)$
$A = 29°$		$z_n = 20.99$	$(>z_p, 0.0001)$	$A = 75°$		$z_n = 53.27$	$(>z_p, 0.0001)$

Appendix Table I. Cont'd (Fig. 11a–c)

Fig. 11a $c = 10$ 0° = north			
No.	%	χ^2	$(>\chi^2; 1\text{ df})$
31	19	42.51	(<0.001)
16	10	3.32	(0.10)
15	9	2.27	
10	6	0.00	
11	7	0.06	
7	4	0.10	
2	1	6.58	
17	10	4.56	(0.05)
5	3	2.64	
6	4	1.72	
6	4	1.72	
8	5	0.47	
2	1	6.58	
1	1	8.29	
3	2	5.07	
23	14	16.11	
$n = 163$	$\chi^2_T =$	102.91	(≪0.001; 15 df)
$r = 0.32$			
$A = 49°$			

Fig. 11b $c = 11$ 0° = north			
No.	%	χ^2	$(>\chi^2; \text{a df})$
23	8	1.47	
9	3	4.41	
10	4	3.47	
7	2	6.62	
7	2	6.62	
1	0	15.93	
6	2	7.89	
14	5	0.84	
42	15	32.56	(<0.001)
26	9	3.69	(0.10)
13	5	1.33	
18	6	0.00	
14	5	0.84	
26	9	3.69	(0.10)
26	9	3.69	(0.10)
44	15	38.18	(<0.001)
$n = 286$	$\chi^2_T =$	131.23	(≪0.001; 15 df)
$r = 0.26$			
$A = \ldots$			

Fig. 11c $c = 17$ 0° = north			
No.	%	χ^2	$(>\chi^2; 1\text{ df})$
22	7	0.27	
21	7	0.09	
24	8	0.94	
32	10	7.70	(0.01)
16	5	0.69	
12	4	3.00	
9	3	5.80	
22	7	0.27	
33	7	0.56	
18	6	0.14	
19	6	0.02	
10	3	4.77	
18	6	0.14	
21	7	0.09	
23	7	0.56	
25	8	1.43	
$n = 315$	$\chi^2_T =$	26.49	(0.05; 15 df)
$r = 0.10$			
$A = \ldots$			

Appendix Table I. Cont'd (Fig. 11d)

Fig. 11d(left) $c = 5$ $0° =$ north				Fig. 11d(right) $c = 5$ $0° =$ north			
No.	%	χ^2	$(>\chi^2$; 1 df)	No.	%	χ^2	$(>\chi^2$; 1 df)
11	20	17.23	(.001)	2	2	2.07	
12	22	22.04	(.001)	2	2	2.07	
11	20	17.23	(.001)	3	4	1.01	
12	22	22.04	(.001)	3	4	1.01	
1	2	1.67		3	4	1.01	
0	0	3.38		5	6	0.02	
0	0	3.38		7	8	0.54	
0	0	3.38		9	11	2.56	
1	2	1.67		18	21	30.30	(<0.001)
0	0	3.38		8	9	1.36	
0	0	3.38		10	12	4.14	(0.05)
0	0	3.38		8	9	1.36	
0	0	3.38		2	2	2.07	
3	6	0.04		3	4	1.01	
3	6	0.04		1	1	3.50	
0	0	3.38		1	1	3.50	
$n = 54$		$\chi^2_T = 108.96$	($\ll$0.001; 15 df)	$n = 85$		$\chi^2_T = 57.49$	(<0.001; 15 df)
$r = 0.78$		$z_c = 3.04$	($>z_p$, 0.05)	$r = 0.48$		$z_c = 1.15$	($<z_p$, 0.05)
$A = 39°$		$z_n = 32.85$	($>z_p$, 0.0001)	$A = 205°$		$z_n = 19.58$	($>z_p$, 0.0001)

Appendix Table I. Cont'd (Fig. 12a–c)

Fig. 12 $c = 9$ (controls, not graphed)				Fig. 12 $c = 9$ (light–dark, not graphed)				Fig. 12 $c = 9$			
No.	%	χ^2	($>\chi^2$; 1 df)	No.	%	χ^2	($>\chi^2$; 1 df)	No.	%	χ^2	($>\chi^2$; 1 df)
18	7	0.25		2	1	11.45		24	8	1.76	
15	6	0.06		2	1	11.45		20	7	0.16	
15	6	0.06		4	2	8.24		12	4	2.18	
16	6	0.00		19	8	0.96		10	3	3.77	
14	5	0.25		34	14	23.30	($<$0.001)	18	6	0.00	
17	6	0.06		29	12	12.56	(0.001)	9	3	4.74	
11	4	1.56		16	7	0.04		17	6	0.09	
14	5	0.25		11	5	1.15		21	7	0.39	
16	6	0.00		3	1	9.78		29	10	6.24	(0.02)
18	7	0.25		9	4	2.52		11	4	2.92	
17	6	0.06		7	3	4.41		26	9	3.22	(0.10)
14	5	0.25		19	8	0.96		22	8	0.74	
25	10	5.06	(0.05)	28	12	10.80	(0.001)	27	0	4.12	(0.05)
18	7	0.25		39	16	37.34	($<$0.001)	10	3	3.77	(0.10)
11	4	1.56		15	6	0.00		17	6	0.09	
17	6	0.06		6	2	5.56		20	7	0.16	
$n = 256$		$\chi^2_T = 10.00$		$n = 243$		$\chi^2_T = 140.54$	($\ll$0.001; 15 df)	$n = 293$		$\chi^2_T = 34.37$	(0.01; 15 df)
$r_d = 0.04$				$r_d = 0.51$				$r_d = 0.08$	$z_c = 0.05$ ($<z_p$, 0.05)		
								$A_d = \ldots$	$z_n = 1.88$ ($<z_p$, 0.05)		

Appendix Table I. Cont'd (Fig. 13a–c)

Fig. 13a $c = 24$				Fig. 13b $c = 6$				Fig. 13c $c = 5$			
No.	%	χ^2	($>\chi^2$; 1 df)	No.	%	χ^2	($>\chi^2$; 1 df)	No.	%	χ^2	($>\chi^2$; 1 df)
85	10	15.60	(0.001)	28	8	2.43		59	24	120.41	($\ll$0.001)
68	8	2.78		19	6	0.17		32	13	17.16	(0.001)
52	6	0.23		22	7	0.06		10	4	2.03	
49	5	0.78		30	9	3.99	(0.01)	13	5	0.44	
79	9	9.89	(0.005)	19	6	0.17		15	6	0.03	
60	7	0.35		13	4	2.97		6	2	5.93	
66	7	1.96		18	5	0.40		6	2	5.93	
39	4	4.94		20	6	0.04		15	6	0.03	
43	5	2.84		17	5	0.72		19	8	0.73	
52	6	0.23		15	5	1.65		17	7	0.12	
47	5	1.32		20	6	0.04		5	2	7.23	
28	3	13.67		14	4	2.26		5	2	7.23	
57	6	0.04		16	5	1.14		4	2	8.65	
58	7	0.11		18	5	0.40		4	2	1.37	
51	6	0.37		27	8	1.80		11	4	1.37	
55	6	0.01		38	11	14.05	(0.001)	29	12	11.45	(0.001)
$n = 889$		$\chi^2_T = 55.11$	(0.001; 15 df)	$n = 334$		$\chi^2_T = 32.28$	(0.01: 15 df)	$n = 250$		$\chi^2_T = 197.36$	($\ll$0.001; 15 df)
$r_a = 0.03$		$z_c = 0.02$ ($<z_p$, 0.05)		$r_a = 0.08$		$z_c = 0.04$ ($<z_p$, 0.05)		$r_a = 0.44$		$z_c = 0.97$ ($<z_p$, 0.05)	
$A_d = \ldots$		$z_n = 0.80$ ($<z_p$, 0.05)		$A_d = \ldots$		$z_n = 2.13$ ($<z_p$, 0.05)		$A_d = 3°, 183°$		$z_n = 48.40$ ($>z_p$, 0.0001)	

REFERENCES

Abele, L., 1970, Semi-terrestrial shrimp (*Merguia rhizophorae*). *Nature* **226**: 661–662.

Altevogt, R., 1955, Some studies on two species of Indian fiddler crabs, *Uca marionis nitidus* (Dana) and *U. annulipes* (Latr.), *J. Bombay Nat. Hist. Soc.* **52**: 702–716.

Altevogt, R., 1957, Untersuchungen zur Biologie und Physiologie indischer Winkerkrabben, *Z. Morphol. Ökol. Tiere* **46**: 1–110.

Altevogt, R., 1964, Lernversuche bei *Uca tangeri*, *Zool. Beitrage* **9**: 447–459.

Altevogt, R., 1965, Lichtkompass- und Landmarken-dressuren bei *Uca tangeri* in Andalusien, *Z. Morphol. Ökol. Tiere* **55**: 641–655.

Altevogt, R., and von Hagen, H. O., 1964, Zur Orientierung von *Uca tangeri* in Freiland, *Z. Morphol. Ökol. Tiere* **53**: 636–656.

Barnwell, F. H., 1965, An angle sense in the orientation of a millipede, *Biol. Bull.* **128**: 33–50.

Barnwell, F. H., 1968, The role of rhythmic systems in the adaptation of fiddler crabs to the intertidal zone, *Am. Zoologist* **8**: 569–583.

Batschelet, E., 1965, Statistical Methods for the Analysis of Problems in Animal Orientation and certain Biological Rhythms, AIBS Monograph, pp. 1–57.

Bliss, D. E., 1968, Transition from water to land in decapod crustaceans, *Am. Zoologist* **8**: 355–392.

Bowers, D., 1964, Natural history of two beach hoppers of the genus *Orchestoidea* (Crustacea: Amphipoda) with reference to their complemental distribution, *Ecology* **45**: 677–696.

Brown, F. A., Jr., 1961, Physiological rhythms, *in* "The Physiology of Crustacea" (T. Waterman, ed.) Vol. 2, pp. 401–430, Academic Press, New York.

Cohen, M., and Dijkgraaf, S., 1961, Mechanoreception, *in* "The Physiology of Crustacea" (T. Waterman, ed.) Vol. 2, pp. 65–108, Academic Press, New York.

Cowles, R. P., 1908, Habits, reactions, and associations in *Ocypoda arenaria*, *Papers Tortugas Lab. Carnegie Inst. Wash.* **2**: 1–41.

Crane, J., 1958. Aspects of social behavior in fiddler crabs, with special reference to *Uca maracoani* (Latreille), *Zoologica* **43**: 113–130.

Datta, L. G., Milstein, S,. and Bitterman, M. E., 1960, Habit reversal in the crab, *J. Comp. Physiol. Psychol.* **53**: 275–278.

Daumer, K., Jander, R., and Waterman, T., 1963, Orientation of the ghost crab *Ocypode* in polarized light, *Z. Vergl. Physiol.* **47**: 56–76.

Edney, E. B., 1968, Transition from water to land in isopod crustaceans, *Am. Zoologist* **8**: 309–326.

Enright, J. T., 1961, Lunar orientation of *Orchestoidea corniculata* Stout (Amphipoda), *Biol. Bull.* **120**: 148–156.

Enright, J. T., 1963, The tidal rhythm of activity of a sand-beach amphipod, *Z. Vergl. Physiol.* **46**: 276–313.

Ercolini, A., 1963*a*, Ricerche sull'orientamento astronomico di *Paederus rubrothoracicus* Goeze (Coleoptera-Staphylinidae), *Monit. Zool. Ital. 70/71*, pp. 416–429.

Ercolini, A., 1963*b*, Ricerche sull'orientamento solare degli Anfipodi. La variazione dell' orientamento in cattivita, *Arch. Zool. Ital.* **48**: 147–179.

Ercolini, A., 1964, Ricerche sull'orientamento astronomico in anfipodi litorali della zona equatoriale. I. L'orientamento solare in una popolazione somala di *Talorchestia martensii* Weber, *Z. Vergl. Physiol.* **49**: 138–171.

Ercolini, A., and Badino, G., 1961, L'orientamento astronomico di *Paederus rubrothoracicus* Goeze (Coleoptera-Staphylinidae), *Boll. Zool.* **28**: 421–432.

Gepetti, L., and Tongiorgi, P., 1967*a*, Nocturnal migrations of *Talitrus saltator* (Montagu) (Crustacea Amphipoda). *Monit. Zool. Ital.* (*N. S.*) **1**: 37–40.

Gepetti, L., and Tongiorgi, P., 1967*b*, Ricerche ecologiche sugli Arthropodi di una spiaggia sabbosa del littorale Tirrenico. II. Le migrazioni di una spiaggia sabbosa del littorale Tirrenico. II. Le migrazioni di *Talitrus saltator* (Montagu) (Crustacea Amphipoda), *Redia* **50**: 309–336.

Gifford, C. A., 1962, Some observations on the general biology of the land crab, *Cardisoma guanhumi* (Latreille), in south Florida, *Biol. Bull.* **123**: 207–223.

Hamner, W. M., Smyth, M., and Molford, E. D., Jr., 1968, Orientation of the sand-beach isopod *Tylos punctatus*, *Anim. Behav.* **16**: 405–409.

Hasler, A. D., 1967, Underwater guideposts for migrating fishes in "Animal Orientation and Navigation," pp. 1–15, Oregon State University Press, Corvallis, Oregon.

Herrnkind, W. F., 1965, Investigations concerning homing, directional orientation and insight in the sand fiddler crab, *Uca pugilator* (Bosc), Unpublished masters thesis, University of Miami.

Herrnkind, W. F., 1966, The ability of young and adult fiddler crabs, *Uca pugilator* (Bosc), to orient to polarized light, *Am. Zoologist* **6(3)**: 298.

Herrnkind, W. F., 1967, The development of celestial orientation during ontogeny in *Uca pugilator*, *Am. Zoologist* **7(4)**: 768–769.

Herrnkind, W. F., 1968*a*, Ecological and ontogenetic aspects of visual orientation in the sand fiddler crab, *Uca pugilator* (Bosc), Doctoral dissertation, University of Miami.

Herrnkind, W. F., 1968*b*, Adaptive visually-directed orientation in *Uca pugilator*, *Am. Zoologist* **8(3)**: 585–598.

Herrnkind, W. F., 1968*c*, Breeding of adult *Uca pugilator* and mass rearing of the larvae with comments on the behavior of the larval and early crab stages, *Crustaceana* (Suppl. 2), pp. 214–224.

Herrnkind, W. F., 1969, Queuing behavior in spiny lobsters, *Science* **164(3886)**: 1425–1427.

Horridge, G. A., 1966*a*, Direct responses of the crab *Carcinus* to the movement of the sun, *J. Exp. Biol.* **44**: 275–283.

Horridge, G. A., 1966*b*, Perception of edge versus area by the crab *Carcinus*, *J. Exp. Biol.* **44**: 247–254.

Horridge, G. A., 1967, Perception of polarization plane, colour and movement by the crab, *Carcinus*, *Z. Vergl. Physiol.* **55**: 207–224.

Hughes, D., 1966, Behavioural and ecological investigations of the crab *Ocypode ceratopthalmus* (Crustacea: Ocypodidae), *J. Zool. Lond.* **150**: 129–143.

Hurley, D. E., 1968, Transition from water to land in amphipod crustaceans, *Am. Zoologist* **8**: 327–353.

Hyman, O. W., 1920, The development of *Gelasimus* after hatching, *J. Morphol.* **33**: 485–523.

Hyman, O. W., 1922, Adventures in the life of a fiddler crab, *Ann. Rep. Smithsonian Inst. 1920*, pp. 443–459.

Jander, R., 1957, Die optische Richtungsorientierung der roten Waldameise (*Formica rufa* L.), *Z. Vergl. Physiol.* **40**: 162–238.

Jander, R., 1963, Insect orientation, *Ann. Rev. Entomol.* **8**: 95–114.

Jander, R., 1965, Die Phylogenie von Orientierungsmechanismen der Arthropoden, *Verh. Deutshc. Zool. Ges. Jena*, pp. 266–306.

Korte, R., 1966, Untersuchungen zum Sehervermogen einiger Dekapoden inbesondere von *Uca tangeri*, *Z. Morphol. Ökol. Tiere* **58**: 1–37.

Linsenmair, K. E., 1965, Optische Signalisierung der Kopulationshohle bei der Reiterkrabbe *Ocypode saratan* Forsk (Decapoda-Brachyura-Ocypodidae), *Naturwissenschaften* **52**: 256–257.

Linsenmair, K. E., 1967, Konstrucktion und Signalfunktion der Sandpyramide der Reiterkrabbe *Ocypode saratan* Forsk (Decapoda Brachyura Ocypodidae), *Z. Tierpsychol.* **24**: 403–456.

Lorenz, K., 1965, "The Evolution and Modification of Behavior," pp. 1–121, University of Chicago Press, Chicago, Ill.

Magni, F., Papi, F., Savely, H. E., and Tongiorgi, P., 1962, Electroretinographic responses to polarized light in the wolf spider *Arctosa variana* C. L. Koch, *Experientia* **18**: 511.

Magni, F., Papi, F., Savely, H. E., and Tongiorgi, P., 1964, Research on the structure and physiology of the eyes of a lycosid spider. II. The role of different pairs of eyes in astronomical orientation, *Arch. Ital. Biol.* **102**: 123–136.

Moore, H. B., Davies, L. T., Fraser, T. H., Gore, R. H., and Lopez, N. R., 1968, Some biomass figures from a tidal flat in Biscayne Bay, *Florida, Bull. Mar. Sci.* **18(2)**: 261–279.

Papi, F., 1955*a*, Orientamento astronomico in alcuni Carabidi, *Atti Soci. Tosc. Sci. Nat. Pisa Mem. Ser. B* **62**: 83–97.

Papi, F., 1955*b*, Astronomische Orientierung bei der Wolfspinne *Arctosa perita* Latr., *Z Vergl. Physiol.* **37**: 230–233.

Papi, F., 1955*c*, Ricerche sull'orientamento astronomico di *Arctosa perita* Latr. (Araneae Lycosidae), *Pubbl. Staz. Zool. Napoli* **27**: 76–103.

Papi, F., 1959, Sull'orientamento astronomico in specie del gen. *Arctosa* (Araneae Lycosidae), *Z. Vergl. Physiol.* **41**: 481–489.

Papi, F., 1960, Orientation by night, the moon, *Cold Spring Harbor Symp. Quant. Biol.* **25**: 475–480.

Papi, F., and Pardi, L., 1953, Ricerche sull'orientamento di *Talitrus saltator* (Montagu) (Crustacea-Amphipoda(. II. Sui fattori che regolano la variazione dell'angolo di orientamento nel corso del giorno; l'orientamento di notte. L'orientamento diurno di altra popolazioni, *Z. Vergl. Physiol.* **35**: 490–518.

Papi, F., and Pardi, L., 1954, La luna come fattore di orientamento degli animali, *Boll. Inst. Mus. zool Univ. Torino* **4**: 1–4.

Papi, F., and Pardi, L., 1959, Nuovi reperti sull'orientamento lunare di *Talitrus saltator* Montagu (Crustacea Amphipoda), *Z. Vergl. Physiol.* **41**: 583–596.

Papi, F., and Pardi, L., 1963, On the lunar orientation of sandhoppers (*Amphipoda, Talitridae*), *Biol. Bull.* **124**: 97–105.

Papi, F., and Serretti, L., 1955, Sull'esistenza di un senso del tempo in *Arctosa perita* (Latr.) (Aranae-Lycosidae), *Atti Soc. Tosc. Sci. Nat. Pisa Mem. Ser. B* **62**: 98–104.

Papi, F., and Syrjamaki, J., 1963, The sun-orientation rhythm of wolf spiders at different latitudes, *Arch. Ital. Biol.* **101**: 59–77.

Papi, F., and Tongiorgi, P., 1963, Innate and learned components in the astronomical orientation of wolf spiders, *Ergebn. Biol.* **26**: 259–280.

Papi, F., Serretti, L., and Parrini, S., 1957, Nuove ricerche sull'orientamento e il senso del tempo di *Arctosa perita* Latr., *Z. Vergl. Physiol.* **39**: 531–561.

Pardi, L., 1954*a*, Esperienze sull'orientamento di *Talitrus saltator*, l'orientamento al sole degli individui a ritmo nicti-emerale invertito durante la "loro notte," *Boll. Inst. Mus. Zool. Univ. Torino* **4**: 127–134.

Pardi, L., 1954*b*, Uber die Orientierung von *Tylos laterillii* (Isopoda terrestria), *Z. Tierpsychol.* **11**: 175–181.

Pardi, L., 1955, L'orientamento diurno di *Tylos laterillii Boll. Inst. Mus. Zool. Univ. Torino* **4**: 167–196.

Pardi, L., 1956, Orientamento solare in un tenebrionide alofilo, *Phaleria provincialls* Fauv. (Coleopt.), *Boll. Inst. Ms. Zool. Univ. Torino* **5**: 1–39.

Pardi, L., 1957, Modificazione sperimentale della direzione di fuga negli anfipodi ad orientamento solare, *Z. Tierpsychol.* **14**: 261–275.

Pardi, L., 1960, Innate components in the solar orientation of littoral amphipods, *Cold Spring Harbor Symp. quant. Biol.* **25**: 395–401.

Pardi, L., and Ercolini, A., 1965, Ricerche sull'orientamento astronomico di anfipodi litorali della zona equatoriale. II. L'orientamento lunare in una popolazione somala di *Talorchestia martensii* Weber, *Z. Vergl. Physiol.* **50**: 225–249.

Pardi, L., and Grassi, M., 1955, Experimental modification of direction-finding in *Talitrus saltator* Montagu and *Talorchestia deshayesei* Aud. (Crustacea-Amphipoda), *Experientia* **11**: 202.

Pardi, L., and Papi, F., 1952, Die Sonne als Kompass bei *Talitrus saltator* (Montagu), Amphipoda, Talitridae, *Naturwissenschaften* **39**: 262–263.

Pardi, L., and Papi, F., 1953, Ricerche sull'orientamento di *Talitrus saltator*. I. L'orientamento durante il giorno in una popolazione del litorale Tirrenico, *Z. Vergl. Physiol.* **35**: 459–489.

Pardi, L., and Papi, F., 1961, Kinetic and tactic responses, *in* "The Physiology of Crustacea" (T. Waterman, ed.) Vol. 2, pp. 365–399, Academic Press, New York.

Pearse, A. S., Humm, H. J., and Wharton, G. W., 1928, Ecology of sand beaches at Beaufort, N. C., *Ecol. Monographs* **12**: 135–190.

Rebach, S., 1968, Orientation and movements of the hermit crab *Pagurus longicarpus*, *Am. Zoologist* **8**: 691.

Salmon, M., and Stout, J. F., 1962, Sexual discrimination and sound production in *Uca pugilator* (Bosc), *Zoologica* **47**: 15–19.

Schöne, H., 1961, Complex behavior, *in* "The Physiology of Crustacea" (T. Waterman, ed.), Vol. 2, pp. 465–520, Academic Press, New York.

Schöne, H., 1963, Menotaktische Orientierung nach polarisiertem und unpolarislertem Licht bei der Mangrovekrabbe *Goniposis*, *Z. Vergl. Physiol.* **46**: 496–514.

Schöne, H., 1965, Release and orientation of behavior and the role of learning as demonstrated in crustacea, *in* "Learning and Associated Phenomena in Invertebretes" (W. H. Thorpe and D. Davenpont, eds.), 135–143, *Animal Behaviour*, Suppl. I, Bailliere, Tindall and Cassell, London.

Schöne, H., and Schöne, H., 1961, Eyestalk movements induced by polarized light in the ghost crab, *Ocypode quadrata*, *Science* **134**: 675–676.

Schwartz, B., and Safir, S. R., 1915, Habit formation in the fiddler crab, *J. Anim. Behav.* **5**: 226–239.

Shaw, S., 1966, Polarized light responses from crab retinula cells, *Nature* **211**: 92–93.

Shaw, S., 1969, Optics of arthropod compound eye, *Science* **165**: 88–90.

Tongiorgi, P., 1959, Effects of the reversal of the rhythm of nycthermal illumination on astronomical orientation and diurnal activity in *Arctosa variana* C. L. Koch (Arenae-Lycosidae) *Arch' Ital. Biol.* **97**: 251–265.

Tongiorgi, P., 1962, Sulle relazioni tra habitat ed orientamento astronomico in alcune specie del gen. Arctosa (Araneae-Lycosidae), *Boll. Zool.* **28**: 683–689.

Tongiorgi, P., 1969, Ricerche ecologiche sugli arhropodi di una spraggia sabbiosa del litorale Tirrenico. III. Migrazioni e ritmo di attivita locomotoria nell'isopode *Tylos latreilli* (Aud. and Sav.) e nei Tenebrionidi *Phalaeria provincialis* Fauv. e *Halammobia pellucida Herbst*, *Redia* **51**: 1–19.

van den Bercken, J., Broekhuizen, S., Ringelberg, J., and Velthuis, H. H. W., 1967, Non-visual orientation in *Talitrus saltator*, **Experientia 23(1)**: 44–45.

von Frisch, K., 1967, "The Dance Language and Orientation of Bees," Harvard University Press, Cambridge, Mass.

von Hagen, H. O., 1962, Freilandstudien zur Sexual- und Fortplanzungsbiologie von *Uca tangeri* in Andalusien, *Z. Morphol. Ökol. Tiere* **51**: 611–725.

von Hagen, H. O., 1967, Nachweis einer kinasthetischen Orientierung bei *Uca rapax*, *Z. Morphol. Ökol. Tiere* **58**: 301–320.

van Tets, G. F., 1956, A study of solar and spatial orientation of *Hemigrapsus oregonensis* (Dana) and *Hemigrapsus nudus* (Dana), B. A. thesis, University of British Columbia.

Waterman, T. H., 1966, Systems analysis and the visual orientation of animals, *Am. Scientist* **54**: 15–45.

Waterman, T. H., and Horch, K. W., 1966, Mechanism of polarized light perception, *Science* **154**: 467–475.

Wenner, A. M., Wells, P. H. and Rohlf, F. J., 1967, An analysis of the waggle dance and recruitment in honey bees, *Physiol. Zool.* **40**: 317–344.

Williamson, D. I., 1951*a*, Studies in the biology of Talitridae (Crustacea, Amphipoda): Effects of atmospheric humidity, *J. Mar. Biol. Assoc. U. K.* **30**: 73–90.

Williamson, D. I., 1951*b*, Studies in the biology of Talitridae (Crustacea, Amphipoda): Visual orientation in *Talitrus saltator*, *J. Mar. Biol. Assoc. U. K.* **30**: 91–99.

Chapter 2

ACOUSTIC SIGNALLING AND DETECTION BY SEMITERRESTRIAL CRABS OF THE FAMILY OCYPODIDAE

Michael Salmon

Department of Zoology
University of Illinois
Champaign, Illinois

and

Kenneth W. Horch

Department of Physiology
University of Utah
Salt Lake City, Utah

I. INTRODUCTION

The burrow-dwelling, semiterrestrial crabs of the family Ocypodidae have radiated to occupy many different types of habitats from the tropics to the temperate regions. In some species, individuals may be relatively isolated from one another, while other species live in dense colonies with adjacent burrows separated by a few inches. The group displays a wide spectrum of difference in behavioral organization and complexity. Only one genus in the family, the fiddler crabs (*Uca*), has received broad study. A summary of current advances knowledge of the behavior, ecology, and physiology of these and other semiterrestrial crustaceans has recently been published in the *American Zoologist* (Barnwell, 1968; Bliss, 1968; Miller and Vernberg, 1968; Salmon and Atsaides, 1968; Schöne, 1968).

The genus *Uca* contains about 100 presently recognized species which, with few exceptions, are found in the intertidal zone of bays and estuaries. Many of these have been studied intensively by Jocelyn Crane over the past 30 years. Crane (1957) divided them into two nonsystematic groupings based upon the form and complexity of the species-typical waving displays show by males. The primitive "narrow fronts" exhibit vertical waves, in which the large claw of the male is lifted and lowered in a vertical plane before the body. The more advanced "broad fronts" show a lateral extension of the claw as well as the vertical component. Intermediates between these two groupings are not uncommon. The two groupings may also be separated geographically. The "narrow fronts" are confined primarily to the Indo-Pacific and the "broad fronts" to Central, South, and North America. Largest concentrations of species in both groups are found in the tropical zones, where they are conspicuous and active during daytime low tides.

The ghost crabs (genus *Ocypode*) consist of 19 species, most of which are found in the Indo-Pacific, although a few occur along the Mediterranean, in West Africa, and along the coasts of the Americas (Bliss, 1968). Only a few have been studied behaviorally in any detail. These are the temperate species, *O. quadrata* (= *albicans*) (Cowles, 1908; Horch and Salmon, 1969), and the tropical forms *O. ceratophthalmus* (Cott, 1929; Barrass, 1963; Hughes, 1966; Horch and Salmon, 1969), *O. laevis* (Fellows, 1966), and *O. saratan* (= *aegyptiaca*) (Magnus, 1960; Linsenmair, 1965, 1967). Burrow location is more variable for some species of *Ocypode* than for *Uca*. Hughes (1966) found *O. ceratophthalmus* occupying intertidal burrows or maintaining holes considerably above the high tide mark. We found similar variability in *O. quadrata* (Horch and Salmon, unpublished observations). *Ocypode* shows several other behavioral differences from *Uca*, e.g., in its preference for beaches directly bordering the ocean, in the proportionately greater distances between adjacent conspecific burrows, and in the tendency for some ghost crabs to show exclusively nocturnal activity.

II. PATTERNS OF ACTIVITY AND DISPLAY

Temperate and tropical populations of fiddler crabs exposed to similar tidal regimes show differences in their daily and tidal patterns of activity. These differences are manifested by shortterm changes in social behavior and the relative importance of visual and acoustic signals mediating these behaviors.

Crane's (1941*a*) earliest studies were primarily confined to the tropical species of fiddlers. These animals live in regions where climatic conditions are suited to year-round reproduction. However, both daily (24-hr) cycles of activity and tidal periodicities considerably modify the time periods when courtship, as manifested by waving activity in males, can occur. For example, the tropical species were only active during diurnal low tides (Crane, 1957). Furthermore, waving activity was apparent only during certain semimonthly diurnal low tides.

Uca maracoani was found to pass through a series of phases in social behavior, each of which characterized the animals' activity for a single low-tide period (Crane, 1958). The least social was the underground phase, during which the animal remained below the surface for at least one complete low tide. This phase was followed by several others, consisting of maintenance and feeding near a burrow, wandering, aggressive wandering, a territorial phase, and finally a display phase when the male waved at nearby females and was sexually receptive. In *U. maracoani*, most of the males reached the display phase when low tides occurred between 0800 and 1000 hr. Other tropical species showed different optimal hours for courtship, with the more primitive species displaying earlier in the day. Since the tides occurred 50 min later each day, males would only court for those few days when low tides coincided with optimal display time, i.e., once every 15 days (semidiurnal tidal pattern).

Barnwell (1968) studied the locomotor activity of other tropical forms (*Uca princeps* and *U. stylifera*) exposed to semidiurnal tidal patterns in Costa Rica. These animals, when placed in aktographs under natural photocycles, showed distinct peaks of activity during times corresponding to low tides in the field. The peaks reached a maximum when the low tides occurred between 0700 and 1200 hr. At this time, the majority of crabs in the field were in their semimonthly display phase. These results suggested that locomotor activity measurements could be used as tentative indicators of courtship activity in fiddler crabs.

In the north temperate zone, reproductive activities of fiddler crabs are restricted to the warmer months. During the winter the adults remain underground in a state of torpor (Crane, 1943). In 1963, North Carolina populations of *Uca pugilator* emerged from their winter burrows in early March but did not begin their waving activity until a month later. By early September, waving activity had ended, even though temperatures were still quite warm (Salmon, 1965). The factors controlling the onset and termination of adult reproductive activity in temperate fiddler crabs remain unknown. It is clear that temperature (Miller and Vernberg, 1968), available food supply (Boolootian *et al.*, 1959), or a combination of both these factors could restrict larval development to the warmer months of the year. Selection pressures upon adults which fail to confine their reproductive efforts to these optimal periods must be considerable.

Uca pugilator, subjected to semidiurnal tidal cycles along the North Atlantic Coast, shows a monthly pattern of courtship distinctly different from the tropical species just described. Waving display is the predominant activity of male crabs during all diurnal low tides (Salmon, 1965). During nocturnal low tides the males produce sound by "rapping" their major cheliped against the substrate (Burkenroad, 1947; Salmon and Stout, 1962). Thus, there is no evidence for a semimonthly rhythm of courtship activity in this species. Rather, the males show almost continuous courtship activity during all low tides, no matter what time of day they occur.

Barnwell (1966) studied the locomotor activity of several temperate species (*Uca pugilator*, *U. minax*, and *U. pugnax*) at the northern extreme of their range (Woods Hole, Mass.). When maintained in aktographs and exposed to natural photoperiods, all three species showed a strong tendency toward higher levels of activity during all low tides, whether they occurred during the day or at night. Both *U. pugilator* and *U. minax* also responded to the daily light–dark cycle, but this 24-hr rhythm showed a different interaction with the tidal (12.4-hr) cycle in the two species. Both species were more active at night than during the day, but in *U. minax* higher nocturnal activity was manifested as a semimonthly periodicity when low tides occurred around the time of sunset (every 15 days). In *U. pugilator*, heightened nocturnal activity in the aktographs always occurred, regardless of the time of low tide.

Recent studies (Salmon, 1967; Salmon and Atsaides, 1968*a*,*b*) have shown that persistent waving activity during diurnal low tides and sound production during some or all nocturnal low tides are consistent components of behavior in eight of the 11 species of fiddlers found along the Gulf and Atlantic Coasts of the United States. The remaining three species (*U. thayeri*, *U. minax*, and *U. subcylindrica*) have not been examined in detail. How might we account for these differences in reproductive behavior between temperate and tropical forms? Our hypothesis (Salmon and Atsaides, 1968*b*) is that persistent courtship activity has been established as a consequence of the shorter periods when climatic and biotic conditions are favorable for reproduction. According to this hypothesis, the heightened seasonal reproductive activity so characteristic of temperate species may actually involve no more time or energy than that devoted to corresponding activities in the tropical forms, with their display phase confined to brief diurnal periods every 15 days.

There is little information available concerning daily patterns of activity in *Ocypode*. To date, only the nocturnal temperate form *O. quadrata* has been tested in aktographs (Harrigan, 1965). Other species have been designated as nocturnal or diurnal on the basis of field observations in restricted areas. The dangers of this approach have been emphasized by Hughes (1966) in his study of *O. ceratophthalmus*. Populations on uninhabited islands or unfrequented beaches were active during the day and night, while those exposed to human disturbance were nocturnally active. Ontogenetic differences are also apparent in *Ocypode*. Juveniles of several nocturnal species have been reported active during the day, perhaps to escape cannibalism by the adults (Crane, 1941*b*; Barrass, 1963; Hughes, 1966). Finally, Linsenmair (1967) has documented apparent racial differences in activity between Egyptian (diurnal) and Ethiopian (nocturnal) populations of *O. saratan*.

Visual and acoustical signals are emitted by *Ocypode*, although their social significance is in most cases unknown. In the tropical species *O. saratan*, the males construct sand pyramids which advertise their sexual phase and attract females to the vicinity of their burrows (Linsenmair, 1965, 1967). Sand pyramids were also observed in colonies of *O. ceratophthalmus* located in Eniwetok, M.I. (Horch and Salmon, 1969). Even though waving display occurs in sexual contexts in other ocypodids (Altevogt, 1957; Griffin, 1965) and in some grapsids (Schöne, 1968), no such displays have been reported in *Ocypode*.

Sound production in ghost crabs was first reported over 100 years ago (see review in Horch and Salmon, 1969). The functional significance of these signals is presently unknown, and only recently have the signals been subjected to physical analysis. The sounds are often produced for brief periods when one crab disturbs another in its burrow. Sounds produced under these circumstances have been interpreted as "territorial," "protest," or "warning" signals (Alcock, 1892, 1902; Cott, 1929; Crane, 1941*b*). Under other circumstances, sounds may be emitted almost continuously during the day or night by aggregations of individuals in closely associated burrows (Barrass, 1963; Horch and Salmon, 1969). Some workers have suggested that this continued "chorus" of acoustically active animals may be related to sexual activities. The only definite connection between sexual behavior and acoustic signalling in the genus was provided by Linsenmair (1967). He observed that female *O. saratan*, orienting to the burrows of males by the sand pyramids nearby, were then attracted into the burrow itself by acoustic signals.

III. SIGNAL CHARACTERISTICS

Fiddler crabs produce sound signals by at least four different methods.

A. Ambulatory Leg Movements

Sound production involving correlated vibratory movements of the ambulatories was first reported by Crane (personal communication), who observed the movements in Trinidad populations of *Uca rapax*. Salmon (1967) confirmed her observations on this species and recorded similar sounds from several closely related forms: *U. pugnax*, *U. burgersi*, and two newly described species, *U. longisignalis* and *U. virens* (Salmon and Atsaides, 1968*b*). The legs show two components of movement during sound production. They are rapidly vibrated, while held off the substrate, at a frequency between 150 and 250 Hz, depending upon species and temperature. They also strike the substrate briefly either before, during, or immediately after vibration, producing a transient "click" with higher-frequency components. Both vibration and striking are usually present in each of the several pulses comprising a single sound (Fig. 1). In some cases, the leg vibration frequency can be estimated by the harmonic intervals shown on sonograms (Fig. 1a,b,e; Watkins, 1967). The pulse duration, number of pulses per sound, and temporal patterning of vibrational and striking components within the pulses differ according to the species (Salmon and Atsaides, 1968*a*).

These sounds have been recorded from male crabs under one of two conditions: during the day, from males just within their burrows after they

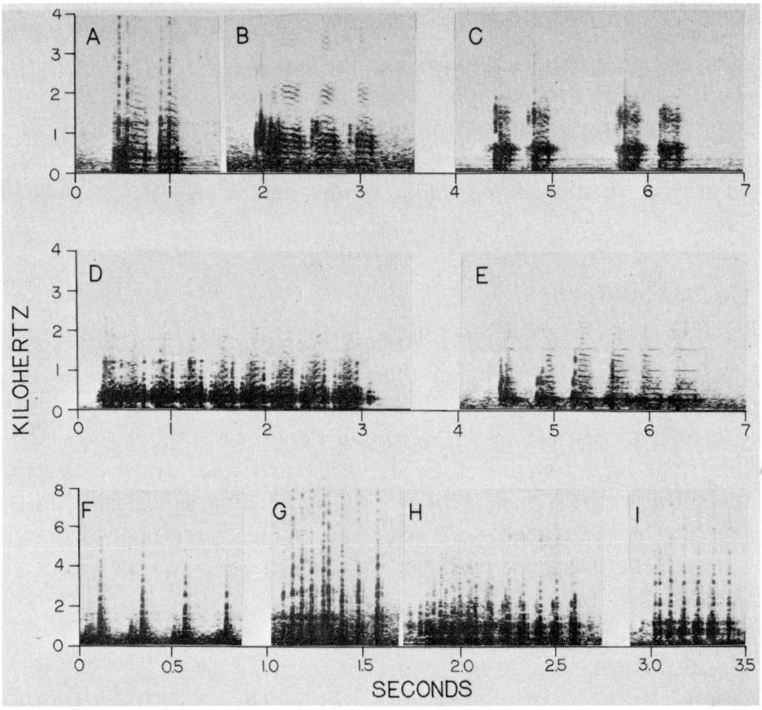

Fig. 1. Fiddler crab sounds produced by leg vibration (top and middle) and rapping (bottom). **(a)** and **(b)** *Uca puguax;* **(c)** *U. virens;* **(d)** *U. longisignalis;* **(e)** *U. rapax;* **(f)** *U. speciosa;* **(g)** and **(h)** *U. spinicarpa;* **(i)** *U. pugilator.* All sounds were recorded with the same type of contact microphone. Sonograms were produced using the narrow filter bandwidth.

have attracted females to their holes by waving; and, at night, from lone males just outside their burrows producing the sounds for prolonged periods (1–3 hr).

B. Rapping

Rapping is the most conspicuous method of sound production used in the genus. The males strike the lower surface of the propodus against the substrate. Each contact between the claw and substrate results in a pulse, and a single sound is composed of two to 11 of these closely spaced pulses (Fig. 1). The sounds have been recorded from three American species (*Uca pugilator*, *U. speciosa*, and *U. spinicarpa*). The spectral energy of the sounds varies, but not according to species; rather, this variation is a function of the substrate (sand, mud, or fine shell) which the claw strikes. Most of the acoustic energy is located below 5 kHz, with a gradual decrease in acoustic energy toward the higher frequencies (8–12 kHz). Each species produces the signals at a species-typical tempo. In *U. spinicarpa*, the rapping tempo is initially highly modulated; i.e., more rapid contacts between the substrate and claw occur during the beginning of the sound.

These sounds are recorded under the same circumstances as those involving ambulatory leg movements in other species—for brief periods during the day from the burrows of sexually stimulated males, and for prolonged periods at night from males producing the signals just outside of their holes.

C. Body Thumping

Female *Uca pugilator* in Florida have been observed to produce sounds by rapidly raising and lowering their bodies within the confines of the burrow. A contact between the body and substrate results in a pulse of sound similar to that produced by contacts between a major chela and the substrate. The sounds were emitted after the crabs were disturbed by scratching the substrate near their burrows or by pushing sand grains into the entrance during the night-time hours. Salmon (1965) erroneously described these movements as being produced by contacts between the minor chela and substrate.

D. Stridulation

Tubercles on the inner surface of the propodus (dactyl base ridges, oblique ridges; Crane, 1967) may be rubbed against ridges on the merus (Dumortier, 1963) to produce stridulatory sounds. Other areas of the body also have been implicated as potential sites for stridulation. Salmon (1965) recorded stridulatory-like sounds from *U. pugnax* and more recently from *U. speciosa* (Salmon, unpublished data). In both cases, the animals involved produced these signals while deep in their burrows, after an "intruder" had been introduced. Consequently, the designation of these sounds as "stridulatory" is tentative, based upon their physical characteristics (high-frequency transients) rather than direct observation.

Crane (1967) has called attention to the potentially significant sounds produced during aggressive interactions between males. These are emitted when one male rubs portions of its major cheliped against specific areas of the claw of a combatant.

The rapping and leg vibrating sounds of fiddler crabs contain air-borne as well as substrate-borne energies. The air-borne components are faint, usually below background levels for a wide-band receiver, even on quiet nights along relatively protected beaches. But the substrate-borne components can be detected a meter or more away with contact microphones (Salmon and Atsaides, 1969).

E. Ghost Crab Sounds

The louder sounds of ghost crabs have recently been subjected to physical analysis (Horch and Salmon, 1969). *Ocypode ceratophthalmus* in Eniwetok, M.I., and Oahu, Hawaii, produces both rapping and rasping (= stridulatory) sounds. Sounds of both types were recorded at night from aggregations of burrows, where they were produced for periods of more than 2 hr. These sounds are probably identical to those described earlier by Hughes (1966) as "knocks" and "rasping" noises. According to Hughes, the knocking signals were very loud, and clearly audible at distances of 20 m. Although Hughes never observed his animals producing the signals, Barrass (1963) observed *O. ceratophthalmus* stridulating just inside the burrow entrance. Sonograms of both sound types (Fig. 2) show similarities in spectral energy distribution, with most of the acoustic energy located below 3 kHz. "Tapping" sounds, recorded exclusively on Oahu, Hawaii, from animals deep within their burrows, have been tentatively attributed to *O. laevis*. The movements involved in the production of this sound await further study. Note, again, that most of the acoustic energy is located at 3 kHz and below.

Ocypode quadrata produced three types of sounds—bubbling, rapping, and rasping. These were recorded during the early morning hours, after the crabs had returned to their burrows from nocturnal foraging trips by the water's edge. Bubbling sounds could be recorded from lone animals in their burrows. After an intruder had been introduced, the sounds would often increase in rate and intensity, and the random bubbling pattern might be modified into a sudden "burst" from one of the two animals (Fig. 2). Rasping sounds were only recorded after one crab was forced into the burrow

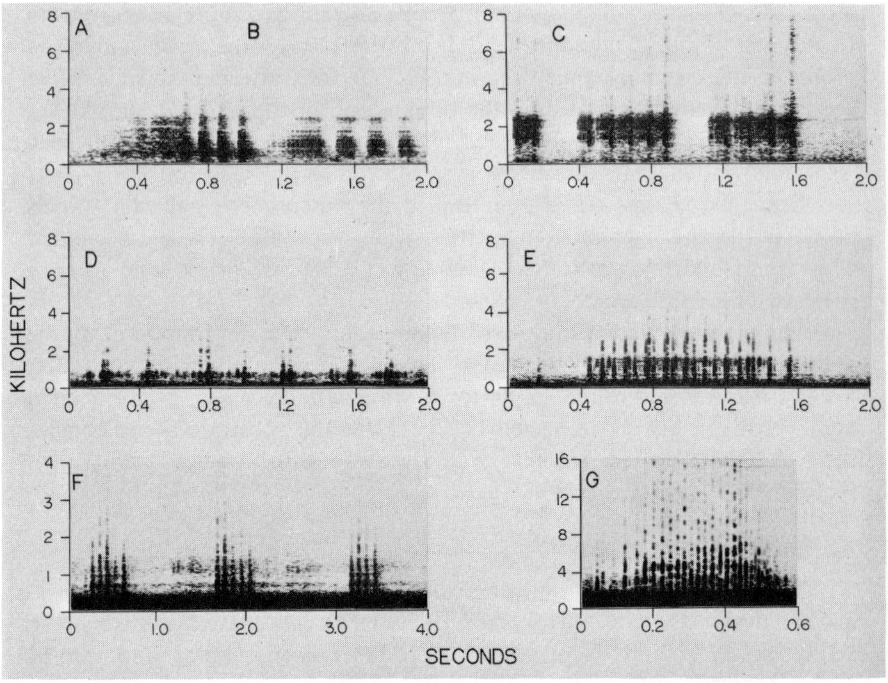

Fig. 2. Ghost crab sounds. Top: Rapping (**a**) and rasping (**b**) sounds recorded simultaneously from a chorus of *Ocypode ceratophthalmus*, Eniwetok, M. I. Note that while the sounds differ in temporal pattern and structure they are similar in spectral energy distribution. (**c**) Tapping sounds, possibly from *O. laevis*, recorded on Oahu, Hawaii. Middle: Bubbling sounds produced in the burrow by *O. quadrata* from North Carolina before (**d**) and after (**e**) an intruder was introduced. Bottom: Rapping (**f**) and rasping (**g**) sounds from the same species.

of a resident. The physical characteristics of this signal suggest that stridulation was involved in its production. Finally, rapping sounds were recorded from four different animals, alone within their burrows. The sounds were only detected immediately after placement of the microphone and therefore may have been elicited by this disturbance.

IV. DETECTION OF TONAL STIMULI

A. Perceptual Capacities

The first attempts to determine sensitivities to acoustic stimuli were carried out exclusively with female *Uca pugilator* by Salmon and Atsaides (1969). Crabs were tested individually by restraining them upon a testing platform mounted above a speaker. Each crab was used once for a test, then discarded. The crab could be viewed from behind a one-way screen, and a 10 sec pause in its escape movements was followed by a 5-sec tonal presentation at a known stimulus intensity. Any movement during the tone presentation was scored as a positive response. By comparing the number of these responses in 20-crab groups of experimental and control (receiving no sound) animals, a measure of sound detection was estimated. Other tests indicated that the crabs responded only to vibration and not to the associated air-borne energies accompanying the stimulus.

Startle response thresholds for frequencies between 30 and 960 Hz are shown in Fig. 3, which also indicates displacement thresholds for vibration "sensitive" and "insensitive" species of insects (Autrum and Schneider, 1948). The crabs showed thresholds comparable to those of the sensitive insects.

Comparisons between *U. pugilator* and *O. quadrata* indicated that the two species differ considerably in their signal detection capacities (Horch and Salmon, 1969). The responses of both species to low (60 Hz) and high (3 kHz) frequencies were studied, again using the startle response technique. Tones were delivered to the subjects in one of two ways: directly through the platform from the speaker below (vibrational component more intense than air-borne) or from a second speaker located some distance away (air-borne component predominant). In either case, vibration amplitudes of the platform and air pressures near the crab were recorded.

At 60 Hz, both species responded only to the vibrational component. Loud, air-borne sounds from the rear speaker failed to produce significant numbers of responses until they yielded platform displacements comparable to the vibratory stimuli. At 3 kHz, *U. pugilator* continued to respond only to platform displacements. But all sound levels of 82 db or greater resulted in significantly more startle responses from experimental *Ocypode* than from their controls. This was the case whether the tones were delivered from the speaker below the platform or from the speaker to the rear, which produced much lower displacements. This finding was especially significant since it was the first experimental demonstration that a crustacean could perceive air-borne sound.

One of us (K.W.H.) has continued these studies with other species of *Ocypode*. A single receptor (Barth's organ) mediates responses to both substrate-borne and air-borne sound. A detailed analysis, including perceptive capacities as well as neural thresholds for intact animals and isolated legs, has been recently published (Horch, 1971).

B. Sound Reception

Horch and Salmon (1969) recorded neural responses to tonal vibrations from the walking legs of *O. quadrata* and *U. pugilator* and presented spectral sensitivity curves for frequencies between 30 Hz and 5 kHz. The values for

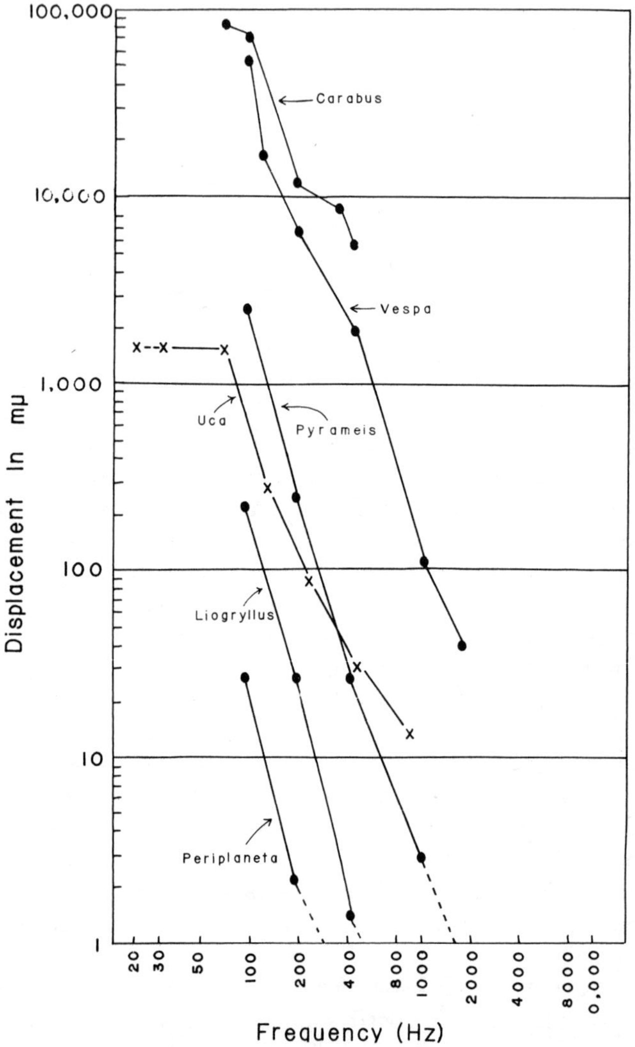

Fig. 3. Substrate vibration thresholds obtained from nerve recordings from the legs of insects (solid circles) and those obtained through unconditioned responses of female fiddler crabs (X). Data for the insects are from Autrum and Schneider (1948). (From Salmon and Atsaides, 1969.)

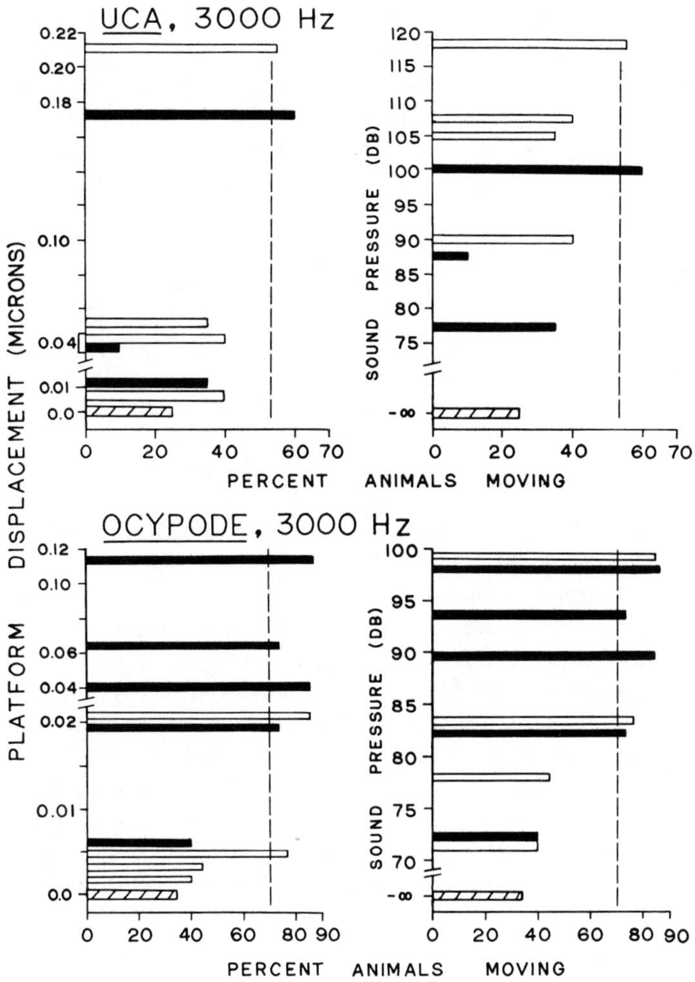

Fig. 4. Percentages of animals showing startle responses (movement) when presented with a 3 kHz tone. Top: *Uca pugilator*. Bottom: *Ocypode quadrata*. Solid bars indicate tones presented directly through the platform; open bars indicate tones from another speaker to the rear. Hatched bars are from controls receiving no stimulus. Each experimental bar represents a group of 17–20 animals; the controls were composed of 40 crabs. Movement percentages are plotted twice to show the platform displacements (left) and sound levels (right, in decibels re 0.0002 microbar) associated with the test stimulus. Dashed vertical lines indicate 0.05 significance level from controls for both species. (From Horch and Salmon, 1969.)

U. pugilator were about a decade *above* those found earlier by Salmon and Atsaides for the comparable frequencies. In order to resolve these disparaties, one of us (M.S.) repeated the electrophysiological recordings on females and expanded the study to include the "honking" species, *U. rapax* (Salmon, 1971).

Pilot tests soon revealed why the results of the two earlier studies differed. Horch and Salmon clamped their animals around the body, allowing them to extend their legs to the test platform. In contrast, Salmon and Atsaides tested subjects standing freely upon the platform. In the latter case, the crabs supported their body weight in a normal fashion, and coupling between the vibration receptors in the legs and the platform was improved. The results from the most recent study, employing freely standing animals, yielded thresholds comparable to those obtained by Salmon and Atsaides. There were also some interesting differences in sensitivity between *U. pugilator* and *U. rapax*. These differences can be best appreciated by examining the acoustic signals of both forms in greater detail.

Spectrographs of single sounds produced by six animals of each species are shown in Fig. 5 and 6 in order to illustrate the variability normally encountered. All of these sounds were recorded with the microphone within 3 cm of the crab.

The "honking" sounds of *U. rapax* (Fig. 5) are composed of two to ten subsounds, each of which contains highly structured frequency components. Each subsound contains maximum energies in narrow frequency bands, beginning at 170–200 Hz. These narrow bands repeat at approximately 180 Hz intervals, up to about 2 kHz in some cases. Sectional displays indicate that most of the acoustic energy is located between 170 and 1500 Hz, with "peaks" between 400 and 1000 Hz. Below 100 Hz, these sounds contain no energies above background levels present under field conditions.

The rapping sounds of *U. pugilator* vary considerably in their spectral energy distribution, especially at the higher frequencies (Fig. 6). Measurable levels of energy are generally present above 8 kHz and extend to the lower frequencies (probably to the acoustic DC), below the range of conventional tape recorders. Most of the energy is concentrated below 1.7 kHz. In some crabs, there is a hint of amplitude structuring within this range. For example, crabs 1, 3, 4, and 6 showed conspicuous peaks between 400 and 800 Hz, and a second peak between 800 and 1500 Hz.

Sensitivity curves for both species are shown for two vibrational parameters, root mean square (RMS) displacement and RMS acceleration, in Fig. 7 and 8, respectively. The values are means $\pm$ standard errors for ten *U. rapax* and 13 *U. pugilator*.

Displacement sensitivities (Fig. 7) tended to increase in both forms with higher frequencies, but not uniformly. Both species were less sensitive to tones of 240 Hz than to the 120 Hz presentations. In *U. pugilator*, there

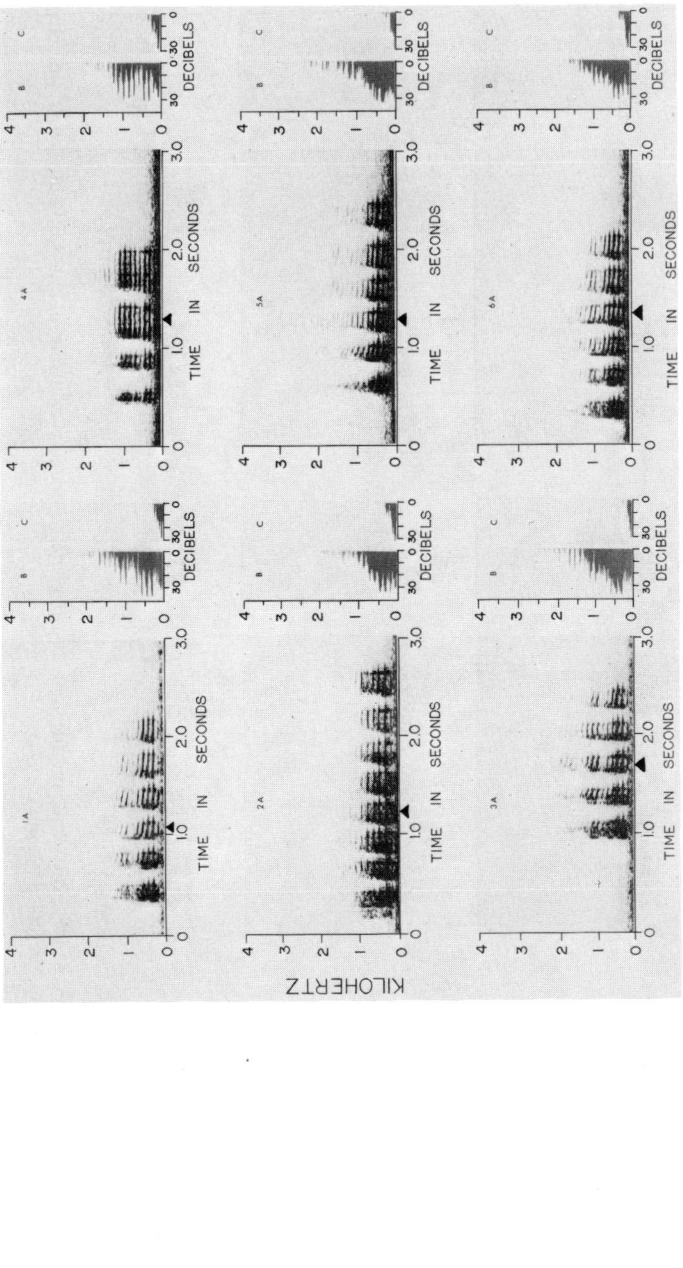

Fig. 5. Sonographic displays of sounds by six *Uca rapax*. **(a)** Frequency vs. time display of the entire sound showing subsound frequency structure and total sound duration. **(b)** Frequency vs. relative amplitude (in decibels) display, as measured at the point in (a) indicated by the triangular mark (below time base). **(c)** Frequency vs. relative amplitude of background noise present just before the sound (Salmon, 1971).

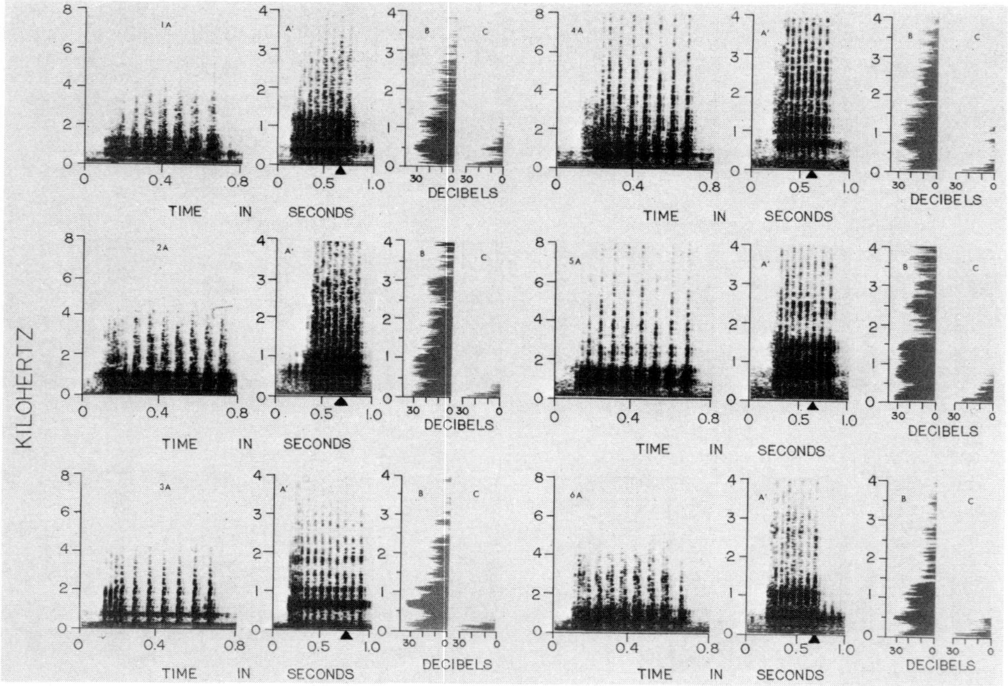

Fig. 6. Sonographic displays of sounds produced by *Uca pugilator*. Format is as in Fig. 5, but two frequency vs. time displays (80–8000 Hz and 40–4000 Hz) are presented (Salmon, 1971).

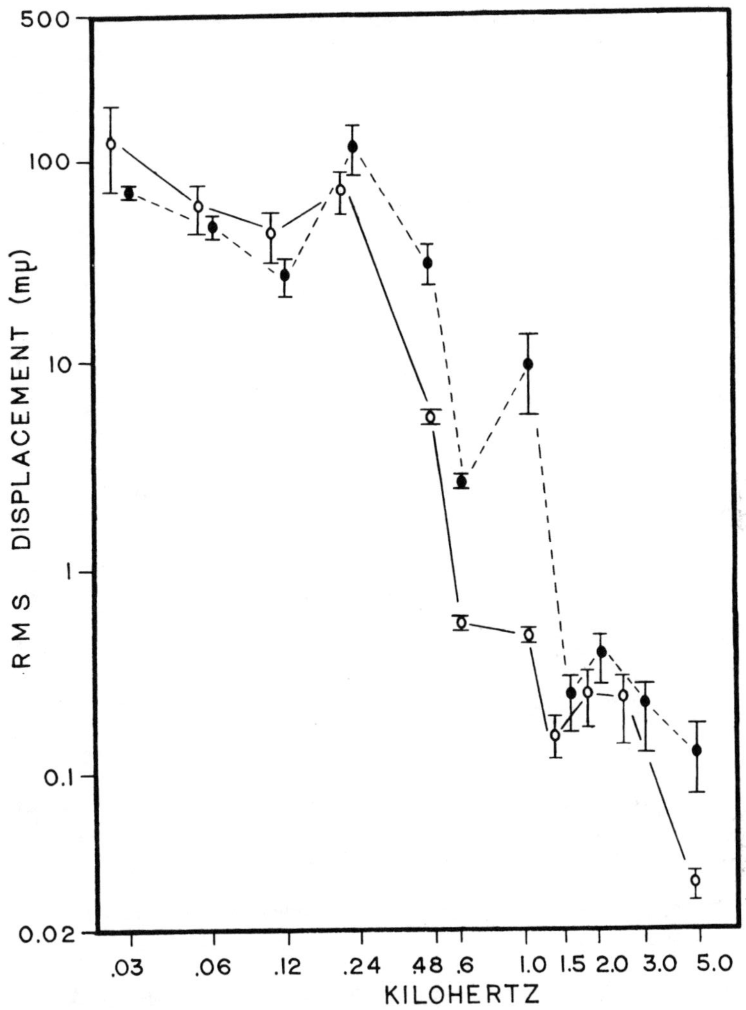

Fig. 7. Vibration response thresholds to tonal stimuli, expressed as RMS displacements. Values represent mean ± standard errors for *Uca pugilator* (solid circles) and *U. rapax* (open circles) (Salmon, 1971).

were sensitivity peaks located at 120 Hz, 600 Hz, and 1.5 kHz. Similar peaks were also evident in our earlier study, although the threshold values for the clamped animals were much higher. In *U. rapax*, the sensitivity was less accentuated or absent. Statistically significant differences between the two forms were present at frequencies of 480, 600, and 1000 Hz, where *U. rapax* was the more sensitive.

The acceleration thresholds calculated from the displacement data

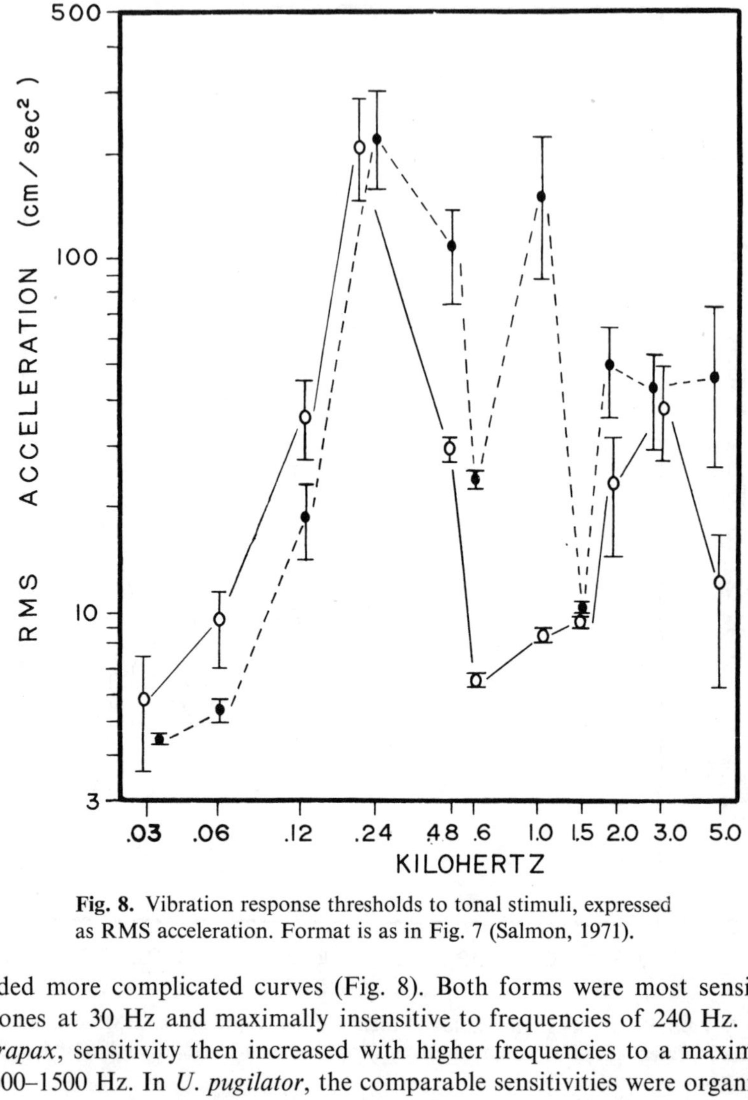

Fig. 8. Vibration response thresholds to tonal stimuli, expressed as RMS acceleration. Format is as in Fig. 7 (Salmon, 1971).

yielded more complicated curves (Fig. 8). Both forms were most sensitive to tones at 30 Hz and maximally insensitive to frequencies of 240 Hz. For *U. rapax*, sensitivity then increased with higher frequencies to a maximum at 600–1500 Hz. In *U. pugilator*, the comparable sensitivities were organized into peaks at 600 Hz and 1.5 kHz. The two forms appeared to differ at 5 kHz, but only a few animals of each species responded to these tones, even at maximal stimulus amplitudes. Significant differences between the species occurred at the same frequencies as those found for displacement. Both forms showed an obvious decrease in variability at threshold value for frequencies where they exhibited maximum sensitivities.

Can these sensitivity curves be related in any way to the spectral energy components of the sounds? The displacement values do not suggest such a relationship, because these values do not take into account energy relationships. For example, a given displacement at a lower frequency can be generated by much lower acoustic energies than the same displacement at a higher frequency. Equivalent accelerations, on the other hand, denote equal acoustic energies regardless of frequency. For this reason, a relationship between signal amplitude structure and receptor sensitivity, if present, should be most apparent from acceleration data (Fig. 8).

The increased sensitivity shown by *U. rapax* to tones between 480 Hz and 1.5 kHz is most obviously related to the spectral energy distributions of their sounds. This range encompasses most of the acoustic energy of the honking sound. The high sensitivity at 30–60 Hz cannot be functional in this regard, since there are no measurable energies in the call below 100 Hz. However, sensitivity at the low frequencies might well be used under other circumstances, such as detection of predators or the movements of conspecifics located near by. These acoustic by-products of movement would undoubtably be composed of low frequency to acoustic DC components.

The same argument could apply to the great sensitivity shown by *U. pugilator* at 30–60 Hz. But, in addition, the rapping sounds of this form contain lower frequencies. Unfortunately, we have not yet measured these, although there is little doubt that a sound generated by contact between two hard objects (compact sand and exoskeleton) would contain these components. The higher-frequency sensitivities at 600 Hz and 1.5 kHz may possibly be related to the amplitude peaks found in some of the sounds. However, much more evidence is needed before the consistency of these peaks can be verified. The sounds of *U. pugilator* do contain considerable energies at these frequencies, and the sensitivity shown by the animals would be useful in signal detection. The crabs apparently "sample" the acoustic spectrum contained in their sounds at low ($<$ 120 Hz), medium (600 Hz), and high (1.5 kHz) frequency bands. Our hypothesis, namely that this system of detection is a mechanism for estimating distance and direction to the signalling crab, will be developed in the next section and elaborated upon in the *Discussion*.

Finally, both forms showed striking insensitivity to frequencies of 240 Hz. Examination of the frequency–amplitude spectrographs (Figs. 5 and 6) show that background noise levels were usually highest between 100 and 300 Hz. Insensitivity to these frequencies could therefore increase the probability of signal detection by improving the signal-to-noise ratio.

V. DETECTION OF RAPPING SOUNDS

A. Characteristics of the Visual–Acoustic System

During diurnal low tides, male *U. pugilator* respond to the approach of a female by increasing their waving rate. If the female comes closer, the male will shift from rapid waving to waving interspersed with rapping and finally to rapping alone just within the burrow entrance. Should the female be enticed to follow the male into his burrow, the rapping sounds will continue until she has descended into the antechamber below, where mating takes place (Crane, 1957).

During nocturnal low tides, males may be observed just outside their burrows producing rapping sounds at relatively slow rates. When touched by another crab, the rapping rate increases and the male simultaneously moves farther into his burrow. If the female is receptive, she follows him and events from here proceed as they do during the day. These male–female interactions are diagrammed in Fig. 9.

Rapid waving by male crabs stimulates neighboring males to increase their waving rates. The rapping sounds of a male also stimulate neighboring males. Salmon (1965) could demonstrate this effect through a series of playback experiments. Daytime tests induced males below the surface to come out their burrows and begin waving. Males already on the surface waved at faster rates during the test. At night, the responses to the playbacks of rapid flurries of sound (typical of stimulated males) were analogous. Males below the surface could be induced to emerge from their burrows and begin emitting sounds. Those already rapping before the test increased their rates of sound production during the playback. These experiments demonstrated that males respond to one another's courtship activities, resulting in the vigorous display activity by groups of males in the vicinity of a female.

Attempts thus far to attract female crabs at night to speakers emitting sounds of conspecific males have been unsuccessful. The problem is a difficult one for serveral reasons, not the least of which is selecting females in an appropriate physiological state. There is at present no reliable way of doing so based upon either external morphology or behavior. However, some aspects of the problem of signal detection by females can be approached by other techniques.

From what distances can a female fiddler crab detect the acoustic signals produced by males? This question can be answered by determining the acoustic energies produced by males, measuring their attenuation with distance, and determining the threshold of females to these sounds. Since rapping sounds are relatively simple noises produced by mechanical move-

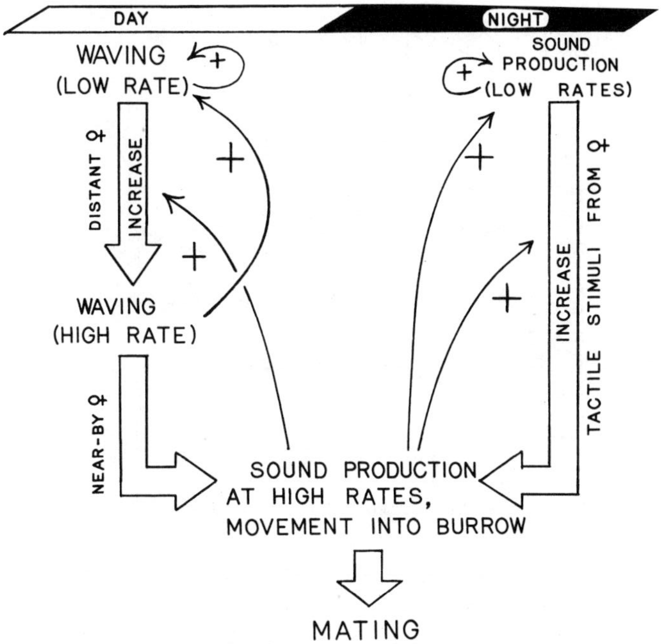

Fig. 9. An outline of courtship sequences in *Uca pugilator*, during the day (left) and at night (right). Open arrows indicate patterns of response by males to stimuli provided by females (outer margins). Thin arrows indicate patterns of response by males to stimuli provided by neighboring males. Plus signs indicate stimulation of the indicated activities in males either not performing them or doing so at low intensities. For further explanation, see text.

ments, an apparatus was designed to simulate these signals under laboratory conditions. The response of females (*U. pugilator*) to these artificial signals was then measured electrophysiologically.

B. Methods

1. Recording Cage and Sand Box

A large Faraday cage with an open front was freely suspended by two wires from pipes near the ceiling (Fig. 10). A sand box, 110 by 37.5 by 9 cm deep, was suspended within the cage and 1 cm above the cage bottom by another pair of wires attached to two different ceiling pipes. Care was taken to assure that there was no contact between the cage and sand box. The suspending wires for the box passed through holes in the top of the cage without touching the screening.

The sand box was lined on the bottom and sides with a single 2.5 cm thick sheet of foam rubber and then filled to the top with a layer of fine

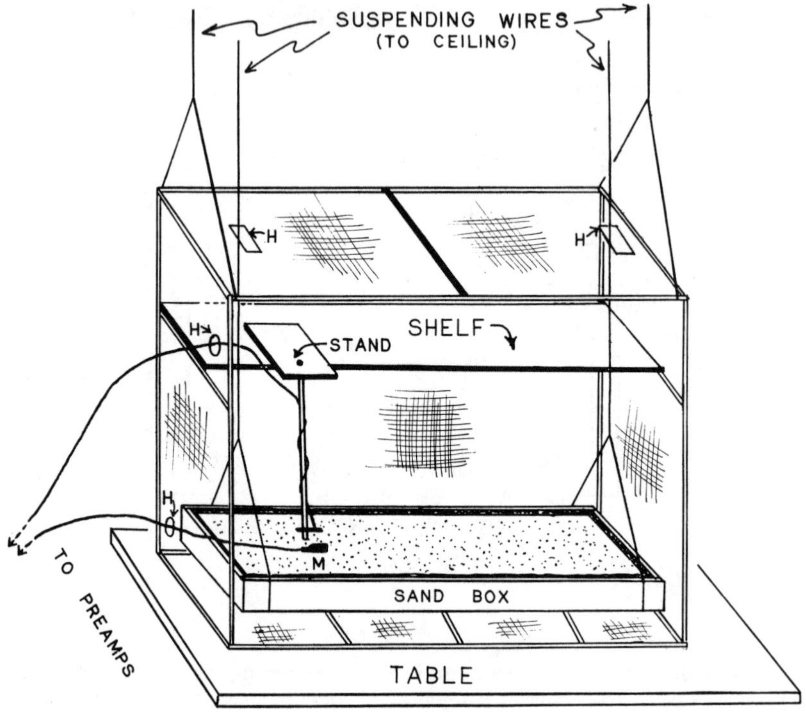

Fig. 10. Diagram of Faraday cage and sand box used during recordings. Each was independently suspended from ceiling pipes so that neither was in contact with the other or the table below. H, holes in cage screening; M, microphone.

sand taken from the surface of a beach in North Carolina where a large colony of crabs was located. It was used unchanged in the box except for the removal of a few pieces of large shell and other debris. The sand was lightly packed into the box and the surface made as level and smooth as possible.

An iron rod, suspended from the frame of the cage, extended to within 3 cm of the sand surface. The recording cable was taped to the bar and extended to a clamp at the bottom. Two No. 32 alligator clips, taped to the clamp, served for attachment of the electrodes, which were composed of fine, insulated silver wire. The recording cable passed out the side of the cage without touching the screen and led to a preamplifier resting upon another table.

A contact microphone (Electro-Voice 805) was placed on the surface of the sand near the clamp and in the long-dimension midline of the box. The cable from the microphone was passed out of the cage, without touching it, to a second preamplifier.

The clamp and microphone were placed about 10 cm from the left side of the sand box. Marks were placed on the right of the sand box at 15 cm intervals, which served as guides for presentations of rapping sounds at various distances from the experimental animals.

2. The "Rapping" Apparatus

A wooden stand was constructed, with its upper surface raised 18 cm above the sand (Fig. 11). The stand was supported by a table under the cage and designed so that it could be moved from left to right within the cage without contacting the screen.

A modified 6 V relay was mounted upon the stand directly above the midline of the sand box. Its contacts were removed and the moving surfaces coated with liquid rubber. A rubber band supplied tension to the armature when no voltage was applied to the relay. Glued to the front surface of the armature was a short, stiff piece of wire. This wire was fastened to a shaft

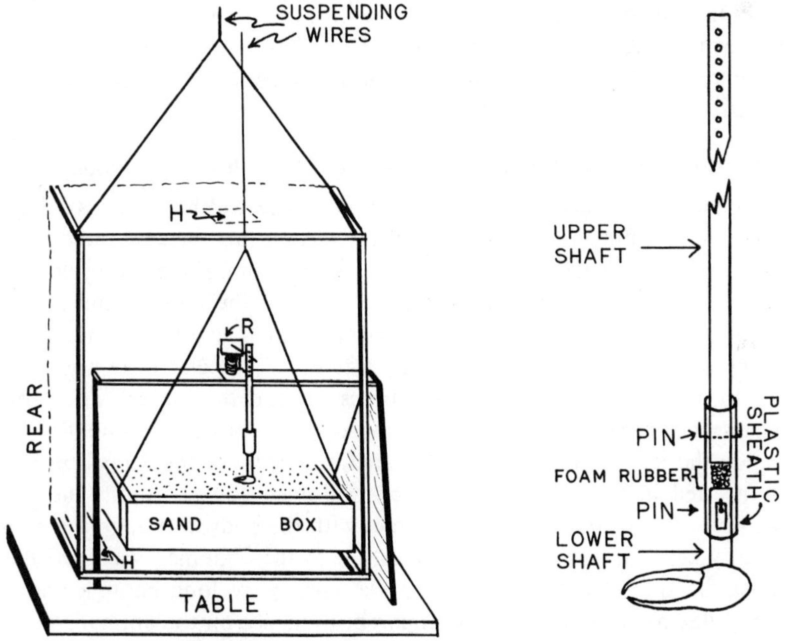

Fig. 11. Left: Side view of cage and sand box, showing position of wood stand with relay (R) and shaft. The stand rested upon the table at the front and through a hole (H) running the length of the cage screening to the rear. Pulsed voltages applied to the relay coil caused the armature to move down and the claw to strike the substrate, one for each pulse. Right: Detailed view of the shaft. The upper shaft contained holes so that the height of the shaft relative to the sand below could be adjusted. A foam rubber wad between the upper and lower shaft acted as a cushion when the claw struck the sand.

made of balsa wood, which contained a male cheliped glued to the bottom (Fig. 11, right).

Rapping sounds were produced by applying pulsed voltages to the relay from a stimulator. Each pulse would cause a rapid lowering of the armature, moving the shaft and its attached claw downward to touch the sand below. Several pulses (three to 15) were delivered to the relay in succession to make one "rapping" sound. The rapping rate, intensity, and total duration of the sound could all be varied by appropriate adjustments to the stimulator. A series of holes in the upper shaft regulated the distance between the claw and substrate. This distance varied from 3 to 6 mm, comparable to the distances observed in the field when males were acoustically active.

The claw shaft was made more elastic by dividing it into two pieces, separated by a small wad of foam rubber. The lower shaft contained a groove through which a pin was fitted. Contact between the base of the claw and the sand caused the lower shaft to move up through the groove and against the foam rubber. This modification served to cushion the claw when it struck the substrate, and retarded the formation of a sand depression. This was true even when sounds were delivered at relatively high amplitudes. As many as 15–29 repetitions of a rapping sound could thereafter be delivered at relatively uniform intensities.

3. Preparation of the Subject; Techniques of Stimulus Presentation

Female *U. pugilator*, 1.2–1.6 cm in carapace width, served as the test animals. Each animal was prepared for tests by cutting the circumesophageal connectives with a small scalpel just below the rostrum. A recording electrode, insulated except at the tip, was inserted through the joint membrane into the merus, just proximal to the meral–carpal joint. The first, second, or third walking leg on either side of the body was used. The electrode was held in place by wrapping the wire several times around the merus. The ground electrode consisted of a small wire noose which passed around the body between the legs. This wire was insulated except for a small portion which was inserted in the space between the abdomen and sternum. The noose was drawn closed to assure good contact with the body, although the animal normally kept the wire pressed in position as the abdomen was held flexed against the sternal plates. The electrode wires were long enough and sufficiently flexible so as to exert no apparent influence on the animal's posture or tonus.

After recovery, the animal was placed upon the sand substrate close to the microphone in the midline of the sand box. The animal would maintain a normal upright posture, remain motionless, and perform normal respiratory movements for up to 54 min. A drop of fresh seawater was given to the animal every 5–10 min to keep its gills moistened. The sand itself was kept moistened by additions of distilled water before each test.

Rapping sounds were produced at distances of 15, 30, 45, 60, and 75 cm from the animal. The presentations were varied so that at each distance the animal received presentations at relatively low and high intensities, and at slow (six to eight beats per second) and fast (ten to 13 beats per second) rapping rates (total of four presentations per animal at each distance). The interval between and duration of the sounds were varied with each presentation. Generally, the sounds were presented every 2–3 sec, with a duration of 0.7–1.5 sec, comparable to the situation in the field.

Several control tests were employed with each animal. In order to be certain that the connectives were cut (and that the animal was not responding to visual stimuli), the claw shaft was elevated so that it moved up and down at the same (or greater) amplitude as in a previous presentation which elicited a clear response. However, the elevated claw did not touch the substrate. As a further precaution, a rapid movement of the hand just above the crab was made at various times during the tests. These hand movements elicited a reflex lowering of the eyestalks but no response from the walking legs if the connectives were severed.

4. Recording and Analysis of the Data

The electrode and microphone signals were preamplified (Tektronix type 122), with filters set to pass frequencies between 8 Hz and 10 kHz. The output of each preamplifier was connected to a channel of a dual-trace oscilloscope and to one channel of a stereo tape recorder. The record level for the stimulus monitor channel was left constant so that all acoustic presentations could be compared. The peak output voltage from the microphone during each presentation, as recorded on the oscilloscope, was noted on tape at the end of a test.

The microphone was later calibrated by simultaneous comparisons between its output in millivolts and decibel readings from a General Radio Co. vibration pickup (type 1560-P-53) and sound level meter. The microphone was found to be flat ± 3 db between 20 Hz and 1.5 kHz and ± 5 db between 20 Hz and 10 kHz. The decibel measurements were then converted to corresponding RMS displacement (millimicrons) or acceleration (cm/sec^2) via conversion tables supplied by the manufacturer.

All the nerve records contained high rates of spontaneous firing. Nevertheless, responses to the rapping sounds could be detected quite easily at higher stimulus intensities, either by visual inspection of the oscilloscope trace or by monitoring the spikes on a loudspeaker. In order to accentuate neural firing in response to weaker acoustic stimuli, the tape recordings were passed through an octave band filter (General Radio Co. 1558-A), with the filter output connected to a speaker. When set to pass frequencies between 300 and 600 Hz or 600 and 1200 Hz, spike burst could be more easily distinguished from the remaining noise.

To obtain a record of this spike activity, the output of the filter was connected to a graphic level recorder, set at a paper speed of 18.75 cm/min and a writing speed of 7.5 cm/sec. This rather slow writing speed had two advantages. First, the records were relatively insensitive to sudden and brief changes in spontaneous nerve activity between successive stimulus presentations. Second, rapid, but brief increases in firing rate in response to each contact between the claw and substrate were summed, since these occurred rapidly and repetitively for every sound. The graphs obtained from the level recorder showed distinct peaks, as a function of time, during periods when a suprathreshold stimulus was presented. These peaks became less distinct at lower stimulus intensities and were absent from presentations which appeared to be below threshold.

5. Field Recordings and Sound Analysis

We wanted to compare the physical properties of the sounds we presented to our animals in the laboratory with those produced by animals in the field. To do so, field recordings were carried out using colonies of crabs located near the Duke University Marine Laboratory, Beaufort, N.C.

In the first series of recordings, carried out with a large colony of crabs in Beaufort Basin, attenuation of the rapping sounds as a function of distance from the animal was measured. An Argonne contact microphone was used to make these measurements. The microphone was placed 2.5 cm from a rapping male and the record level of the tape recorder (Uher 4000 L, tape speed 3.75 ips) adjusted so that the sounds peaked at −2 to −3 db on the level meter. Recordings were repeated, without changing the record level, at distances of 5, 10, and 20 cm between the microphone and crab. The male was left undisturbed throughout these recordings. After they were completed, the male was captured and his carapace width measured.

A second series of recordings was carried out using another colony of crabs located on a beach in front of the laboratory. This colony was small but near to a power source for a preamplifier, oscilloscope, and tape recorder. The Electro-Voice microphone used in laboratory presentations was employed for these recordings. It was placed 2.5–3.0 cm from each of 11 males while they were acoustically active. Microphone output voltages, as displayed on the oscilloscope, were recorded while the sounds were stored on tape. These data were later converted to the corresponding acceleration value, in cm/sec^2.

With these two sets of recordings, we could compare (1) relative attenuation of the sounds in the field with those measured in the laboratory and

(2) sound intensities produced by animals in the field with those delivered to our test crabs under artificial conditions. In addition, we could make some rough comparisons between the frequency spectra of the rapping sounds produced in the laboratory and those of males in the field, especially with reference to attenuation of certain frequencies with distance.

C. Results

1. Characteristics of Rapping Sounds Under Laboratory and Field Conditions

Sonograms of sounds presented to female crabs in the laboratory, as a function of distance and rapping rate, are shown in Fig. 12. These fall close to the mean for amplitudes delivered at each distance.

At 15 cm, the sounds contained frequencies above 8 kHz, but with most of the acoustic energy confined to frequencies below 4 kHz. As the distance between the rapping apparatus and the animal was increased, the higher frequencies were attenuated. At 40 cm between the animal and rapping apparatus, most of the acoustic energy was confined to 3 kHz and below.

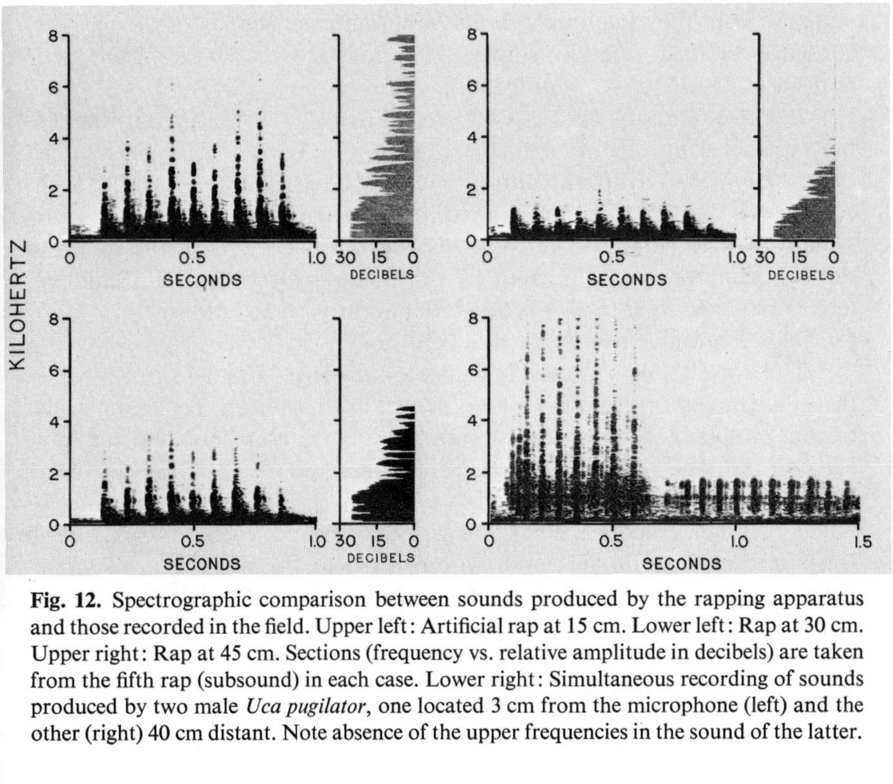

Fig. 12. Spectrographic comparison between sounds produced by the rapping apparatus and those recorded in the field. Upper left: Artificial rap at 15 cm. Lower left: Rap at 30 cm. Upper right: Rap at 45 cm. Sections (frequency vs. relative amplitude in decibels) are taken from the fifth rap (subsound) in each case. Lower right: Simultaneous recording of sounds produced by two male *Uca pugilator*, one located 3 cm from the microphone (left) and the other (right) 40 cm distant. Note absence of the upper frequencies in the sound of the latter.

A field recording is also shown in Fig. 12 (lower right). Frequency attenuation patterns under field conditions were similar to those observed in the laboratory. The sonogram illustrates two sounds, recorded concurrently, produced by a male 3 cm from the microphone and another 40 cm distant. Note that frequencies above 2 kHz are absent from the latter.

While the frequency spectra and rapping tempos of artifical and natural sounds were similar, a comparison between the two (Figs. 6 and 12) showed that differences were evident. Most pronounced were the greater subsound durations at 800 Hz and below in the laboratory presentations, probably a result of reflection of low-frequency sound from the walls and bottom of the sand box. Also, spectral energies below 1 kHz tended to be more uniform in amplitude distribution within the laboratory signals.

Amplitudes of artificial rapping sounds used for laboratory presentations were recorded in terms of microphone output voltages next to the test crab. A plot of all amplitudes is shown in Fig. 13. Each point represents a single presentation. The most intense presentations were made when the claw struck the substrate closest to the crab (15 cm), but a good deal of overlap was evident. At 75 cm, microphone output varied from a high of 2.5 mV to a low of 0.33 ŋV, the weakest sound amplitude we could present. Output voltages from the same microphone, resulting from sounds of 11 males recorded in the field, are also shown. They ranged from 1.1 to 9.7 mV, with a mean of 4.6 mV 2.5–3.0 cm from the male.

Relative rates of sound attenuation in the field, as measured by the output voltages from tape recordings, are shown in Table I. The mean values ranged from 3.6 V from the output of the tape recorder at 2.5 cm from the crab to 1.6 V at 20 cm from the crab. Attenuation rates for laboratory presentations were compared to those shown in Table I. When the slopes of the two samples were compared by regression analysis, no significant difference was apparent ($b = -0.238$ for laboratory presentations, $b = -0.263$ for field recordings, $t = 0.143$, $p > 0.50$).

The correlation between animal size and level of sound produced in the substrate 2.5 cm from the crab is shown in Fig. 14. The peak amplitudes from a sample of five or more sounds per animal, all recorded at the same gain on the tape recorder, were used for these points. A trend for larger animals to produce louder sounds was clearly evident. It should also be noted that fiddler crab sounds are generally below levels of background when measured in the air, even within 2 cm of the male. But substrate-bone sounds may be as much as 25 db above background. Clearly, the use of the substrate as a channel is optimal for the crab, given its mode of signal production.

2. Neural Responses of Females to the Sounds

The responses of 16 females, all but two tested at each distance, are shown in Fig. 13. Each crab responded to the presentations at 15 cm. Since

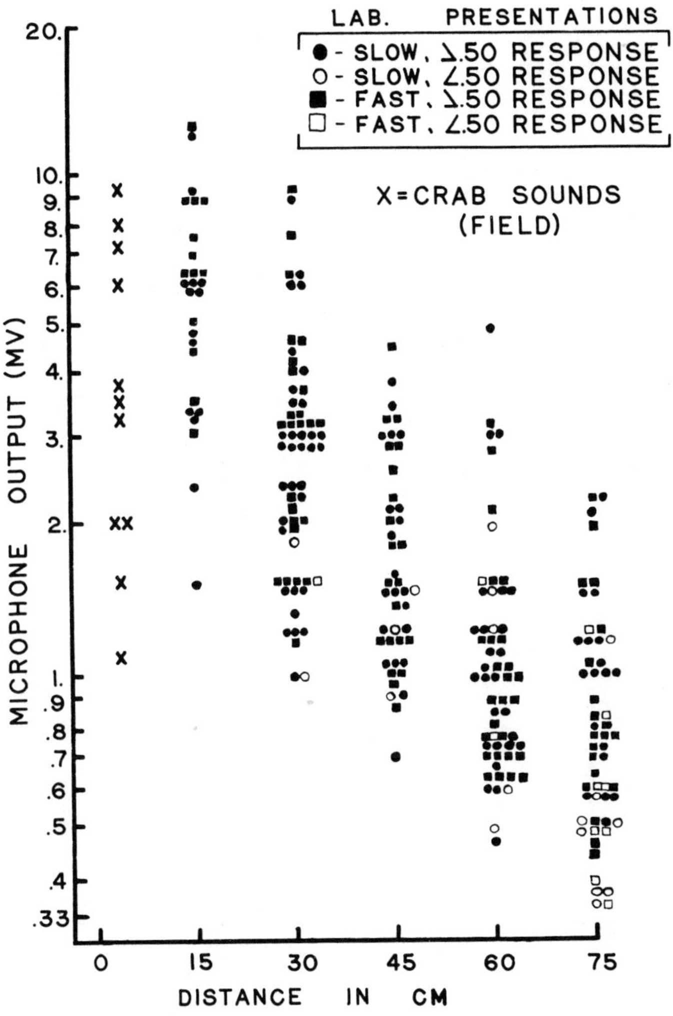

Fig. 13. Responses of female crabs to artificial rapping sounds presented at 15–75 cm distant and differing in amplitude (as measured by microphone output voltages near the crab) and rapping rate. Each symbol indicates which presentation (four per crab, 30–75 cm) was above threshold. X, amplitude of sound produced by 11 male crabs in the field, as measured by output voltages from the same microphone 2.5 cm from the male.

Table I. Attenuation of Rapping Sounds by *Uca pugilator*, as a Function of Distance[a]

Distance (cm) between crab and microphone	Number of crabs	Tape recorder output (V)		
		Range	Mean	SD
2.5	21	3.0–4.0	3.6	0.44
5.0	17	1.7–4.0	2.9	0.74
10.0	21	1.0–3.8	2.5	0.81
20.0	20	0.6–3.4	1.6	0.83

[a] Gain on the tape recorder was selected to give approximately equal readings at 2.5 cm, then left constant.

it became apparent early in the study that all tests at this distance would be above threshold, only one presentation at each rapping rate was given to the last seven animals at this distance. This accounts for the fewer points on the graph at 15 cm.

At 30 and 45 cm, only three presentations failed to elicit responses by the test crabs. These were due to slight changes in the position of the leg which sometimes occurred just before or during the presentation. Virtually all positional changes had a profound effect on the sensitivity of the crabs. For microphone outputs of 0.6 mV and above, we could obtain responses

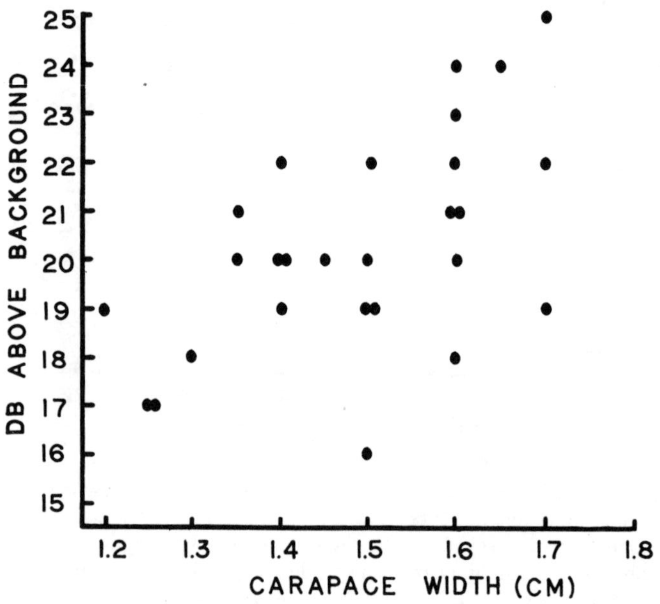

Fig. 14. The correlation between crab size (carapace width) and relative intensity of rapping sounds, as recorded at 2.5 cm from the male. All sounds were recorded on a single night.

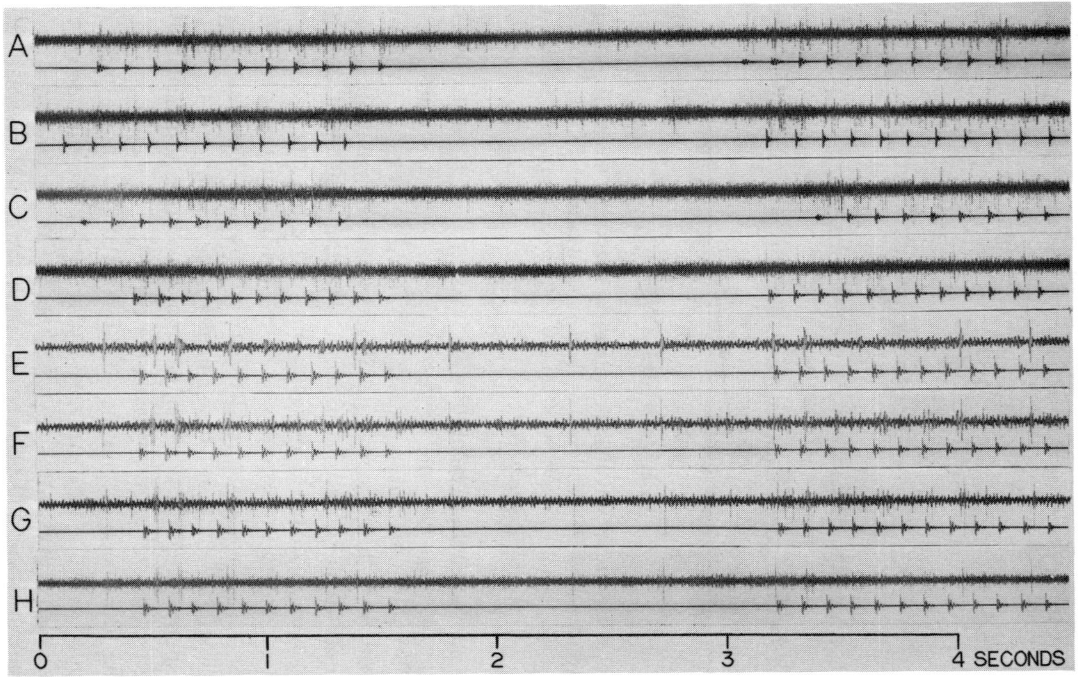

Fig. 15. Filtered and unfiltered nerve records of responses by a crab to stimulus presentations. All of these were above threshold. Upper trace: Nerve record. Lower trace: Rapping stimulus. Unfiltered nerve records: (**a**) 15 cm between stimulus and crab; (**b**) 30 cm; (**c**) 45 cm; (**d**) 60 cm. Filtered nerve records of the presentation at 60 cm: (**e**) passing only frequencies between 150 and 300 Hz; (**f**) between 300 and 600 Hz; (**g**) between 600 and 1200 Hz; (**h**) between 1.2 and 2.4 kHz. Note that filtering accentuates the clarity of the responses, especially in (f) and (g).

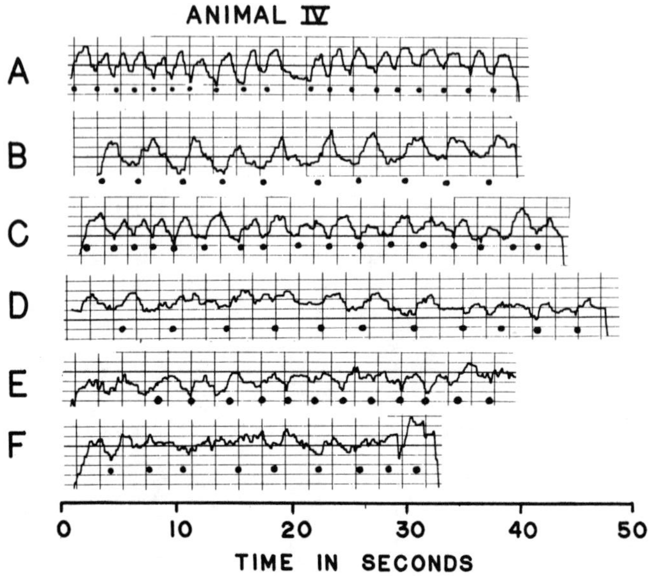

Fig. 16. Graphic level recorder records of filtered nerve responses (only 300–600 Hz passed) by crab 4. The onset of each sound is indicated by the dots below each record. Distance and microphone output: **(a)** 15 cm, 12.5 mV; **(b)** 30 cm, 4.0 mV; **(c)** 45 cm, 3.0 mV; **(d)** 60 cm, 1.4 mV; **(e)** 75 cm, 1.0 mV; **(f)** 75 cm, 0.5 mV.

to a previously subthreshold presentation by coaxing the animal to adjust its leg position again. This was not the case for less intense presentations.

At lower stimulus amplitudes, the animals failed to respond to some or all of the sounds in the presentation. Six different animals were tested with presentations which ranged in microphone output between 0.42 and 0.53 mV. These tests involved 11 presentations (two at 60 cm, nine at 75 cm) in which five resulted in responses to more than half of the sounds in the entire presentation, while six resulted in responses to less than half. Five lower levels of presentation (0.33–0.4 mV) to three animals failed to produce any responses, regardless of leg position (two crabs tested at both fast and slow rates, one crab at a slow rate; see Fig. 13).

Filtering the nerve records considerably improved the signal-to-noise ratio and provided a more sensitive measure of thresholds. Examples of filtered and unfiltered signals are shown in Fig. 15. When all frequencies but those between 300 and 600 Hz were rejected, the neural responses to rapping stimuli were most easily detected, either visually (inspection of the oscillograph records) or by acoustic measurements. In none of the records was any indication of a mechanical stimulus artifact evident.

An example of a filtered record, as analyzed by the level recorder, is shown in Fig. 16. This animal showed obvious responses to presentations at 15, 30, and 45 cm. A movement of the walking leg just before the presentation at 60 cm reduced the sensitivity of the preparation, but responses to the majority of the sounds were still evident. The two lower records show responses to two different sound amplitudes at 75 cm. The higher amplitude (1.0 mV) was clearly above threshold, while the lower (0.5 mV) was designated as below threshold.

Figure 17 summarizes the responses of all test crabs under laboratory

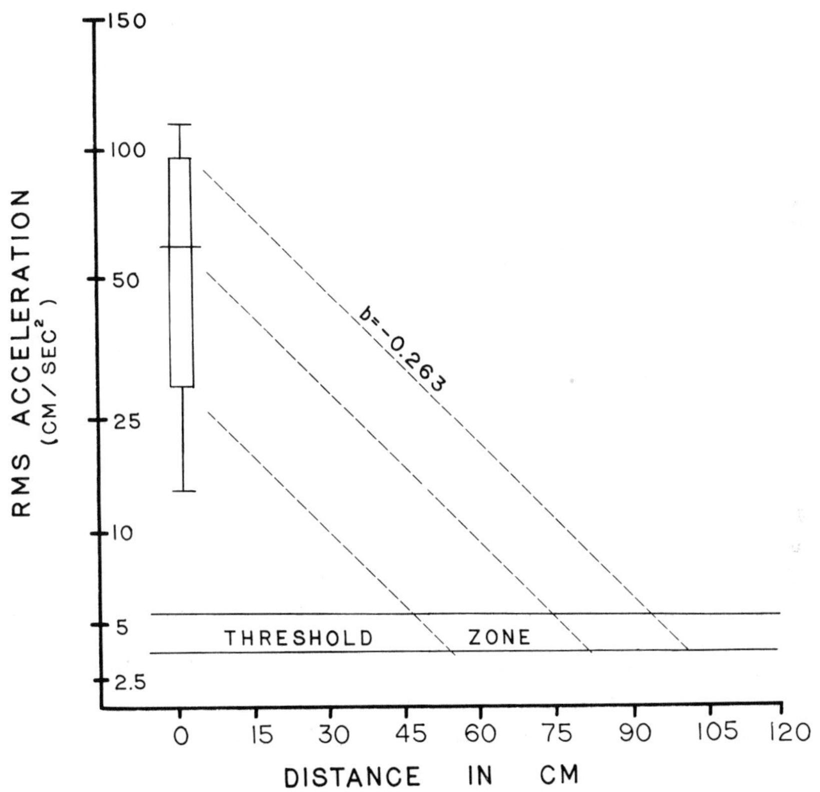

Fig. 17. Calculated distances at which the sounds of male *Uca pugilator* could be detected neurophysiologically by conspecific females. The zone between 3 and 6 cm/sec² acceleration corresponds to the threshold presentations to the females (microphone output of 0.42–0.53 mV, Fig. 13). At the left is the mean (horizontal line, 62 cm/sec²), standard deviation (open vertical bar), and range (vertical bracket) of acoustic energies produced by 11 males in the field (see Fig. 13 for individual measurements). Finely dashed lines indicate distances at which the acoustic energies at the extremes of the standard deviation, as well as the mean, would reach threshold for detection by females. The slope of these lines ($b = -0.263$) is based upon attenuation measurements given in Table I.

conditions and indicates acoustic components in terms of RMS acceleration. In terms of our female crabs, presentations between 3–6 cm/sec^2 acceleration (corresponding to 0.42–0.53 mV from the microphone; see Fig. 13) resulted in about half of the animals responding. This range of stimulus amplitudes is therefore designated as the threshold. The brackets, vertical bar, and horizontal line indicate the range, standard deviation, and mean of accelerations produced by 11 males in the field, as recorded at a distance of 2.5 cm. The finely dashed lines show the rate of attenuation of these sounds with distance in relation to the threshold measurements for females. The slope is calculated from levels recorded in the field at 2.5–20.0 cm (Table I).

For the lower sound amplitudes (SD = 30 cm/sec^2 acceleration), thresholds for detection would be reached at a distance of about 30 cm; for sounds at the higher SD (94 cm/sec^2), the threshold would be reached at about 100 cm between the female and the sound source.

VI. DISCUSSION AND CONCLUSIONS

Our experiments and measurements have demonstrated that female crabs will show neural responses to artifical sounds mimicking those of conspecific males. From our calculations based upon rates of decay of crab sounds and the actual levels of these signals produced under field conditions, female *U. pugilator* should, on the average, be capable of detecting males at least 75 cm away. They are potentially capable of detecting them at distances in excess of 100 cm. Larger males, which produce louder acoustic signals, have greater calling ranges.

These predictions are possible because sounds recorded in the field have similar frequency spectra at similar intensities. However, certain assumptions are made. The recipient crab must maintain a leg posture which makes it as sensitive to vibration as our laboratory subjects. Background noise levels in the field are ignored, though they must decrease the threshold distance. Finally, it is assumed that the acoustic impedance of the substrate for rapping sounds under most field conditions is comparable to the one we have sampled (Table I).

There is little doubt that our sample of 11 males did not reach the upper limit of sound amplitude for the species. When the size of fiddler crabs and the proximity of their burrows in the colony are considered, the potential "calling range" of the male is relatively large. The estimates are also consistent with observations on the distances separating chorusing males when they respond to one another's rapping sounds.

Sound production by fiddler crabs has apparently evolved independently within the genus. One group of closely related species (group II; Crane,

1941*a*) produces the honking sound, correlated with rapid and complex movements of the ambulatories. Other species, belonging to several of Crane's other groups, produce sound by rapping. Only one species, *U. thayeri*, is presently known to emit signals of both types (von Hagen, personal communication).

The movement patterns from which these acoustic signals evolved are presently unknown. Some possibilities may be suggested. The "rapping" species have in common a rapid waving gesture, in which the claw is usually brought down toward the substrate between successive waves in a rapid fashion. In several tropical forms, Crane found that the claw actually strikes the substrate as each wave is terminated, sometimes repeatedly, before the claw is again elevated for the next wave. The emancipation of these striking movements from the context of waving, to the confines of the burrow entrance and the burrow itself, probably occurred gradually. Presently, a number of tropical species and most (if not all) of the temperate rapping forms produce these signals within the burrow in response to approaches by females.

We suggest that the honking sound might be derived from another movement pattern, one not directly associated with waving. All fiddler crabs perform rapid leg-flicking movements within the burrow. These movements are elicited by intruders and probably aid the resident in spatial localization of the intruder by tactile cues. Leg flicking, when monitored with a contact microphone, is detected as discrete bursts of sound as several legs are simultaneously extended and swept downward to the substrate. In the group II crabs, the flicking movements by a male often grade into honking sounds as a resident male makes tactile contact with a female within the burrow. More specifically, in *U. rapax* the leg-flicking sound bursts are produced at about the same rate as the subsounds of the species, which further suggests a close association between the two behaviors.

Our threshold data show that sound signals are detected by female *U. rapax* and *U. pugilator* with considerable sensitivity. Both species are exposed to similar ambient noise spectra in the field, which could accouut for their parallel insensitivity to frequencies around 240 Hz. In *U. rapax*, sensitivity to higher frequencies suggests tuning of the vibration receptors to the maximum energies of the honking sound. The amplitude structure of the signal warrants a broad sensitivity to its several narrow frequency bands, which vary from one male to the next as well as with temperature.

The sensitivity curves for *U. pugilator* suggest that the rapping sounds are detected by sampling the signal at three bands of frequencies. An examination of the sensitivity curves in relation to the amplitude of the signals is particularly instructive here. At 240 Hz and 1 kHz, the mean thresholds were above 100 cm/sec^2 acceleration. At a distance of 2.5 cm from the male, the *total* acoustic amplitude of the rapping sounds was generally well below that value. It follows that signal detection, especially at a distance, cannot involve these frequencies. In effect, the high thresholds at 240 Hz and 1 kHz are effective filtering mechanisms for the rapping signal.

Some other details of signal detection in *U. pugilator* emerge from our data. The thresholds of females to the artificial rapping sounds spanned a range of 3–6 cm/sec^2 acceleration (Fig. 17). At this amplitude, the frequencies contained in the sounds were confined to 1.5 kHz and below. We can conclude that at these stimulus levels the responses must be mediated by the low-frequency detectors (30–60 Hz), since only their thresholds were sufficient for signal detection (mean of 4.2 cm/sec^2 acceleration).

Why is *U. pugilator* also sensitive to frequencies of 600 Hz and 1.5 kHz, when its thresholds at lower frequencies would be sufficient for signal detection at greater distances? Our hypothesis is that sensitivity to the two higher-frequency bands provides information for estimating distance from and direction to the sound-producing male. A female approaching a male from a distance would first detect only the lowest frequencies contained in his sounds. Continued movement toward the male would result in stimulation of both the low- and intermediate-frequency (600 Hz) detectors. Finally, as the female came within a few centimeters of the male, the high-frequency detector would also be stimulated. These predictions follow by considering the relationships between the sensitivities of the receptors to the three bands of frequencies, the amplitude and spectral energy distribution of the rapping sound, and the attenuation patterns of these frequencies with distance.

Similar patterns of frequency change may also accompany attenuation of honking sounds as they are propagated through the substrate. Distance to male might be estimated in a parallel manner by females of *U. rapax*. But directional localization could be accomplished by other mechanisms. Horch (1971) has found "phase locking" of spikes from the vibration receptors of *Ocypode* to low-frequency (up to 600 Hz) vibrations. The extended duration of the subsounds of *U. rapax*, as well as the frequency structuring of its signal, suggests that its neural responses to these sounds would be similar to the phase locking shown by *Ocypode*. Bilateral comparisons of phase relationships between neural responses of legs nearest and farthest from the sound source could provide the directional information. But all of these speculations await verification by experiment.

It is obvious from this review that the fiddler and ghost crabs present us with unique systems for analysis of communication. Substrate-borne sound is utilized in a fashion quite analogous to air-borne acoustic signals in other invertebrates and vertebrates. Our studies to date have only begun to define the characteristics of the system in terms of signal propagation, receptor sensitivity, cues in the signals carrying information, and other important parameters. Even at this time there is no quantitative evidence that the acoustic signals (or their waving display counterparts in *Uca*) actually function as courtship and/or agonistic signals. The significance of air-borne sound detection by *Ocypode* is also unknown. The possibilities for future experimental work are virtually unlimited.

ACKNOWLEDGMENTS

We are grateful to Mr. Gary W. Hyatt, Mr. Daniel Harms, and Mr. John Frey for assisting in the field work and data analysis. Drs. Anne Sayler, Arthur N. Popper, and Brian A. Hazlett criticized the manuscript. Dr. Sayler carried out most of the statistical analysis. The staff of the Duke University Marine Laboratory and Mote Marine Laboratory cooperated in many ways to facilitate this research. The studies outlined here were supported by funds from the University of Illinois Research Board, and by grants from the National Science Foundation.

This paper is a contribution of the Mote Marine Laboratory and the Duke University Marine Laboratory.

REFERENCES

Alcock, A., 1892, On the stridulating apparatus of the red ocypod crab, *Ann. Mag. Nat. Hist.* **6(10)**: 336–337.

Alcock, A., 1902, "A Naturalist in Indian Seas," London.

Altevogt, R., 1957, Beitrage zur Biologie und Ethologie von *Dotilla blanfordi* Alcock und *Dotilla myctiroides* Milne-Edwards (Crustacea Decapoda), *Z. Morphol. Ökol. Tiere* **46**: 369–388.

Autrum, H., and Schneider, H., 1948, Vergleichende Untersuchungen über den Erschütterungssinn der Insekten, *Z. Vergl. Physiol.* **31**: 77–82.

Barnwell, F. H., 1966, Daily and tidal patterns of activity in individual fiddler crabs (genus *Uca*) from the Woods Hole region, *Biol. Bull.* **130**: 1–17.

Barnwell, F. H., 1968, The role of rhythmic systems in the adaptation of fiddler crabs to the intertidal zone, *Am. Zoologist* **8**: 569–583.

Barrass, R., 1963, The burrows of *Ocypode ceratophthalmus* (Pallas) (Crustacea, Ocypodidae) on a tidal wave beach at Inhaca Island, Mocamique, *J. Anim. Ecol.* **32**: 73–85.

Bliss, D. E., 1968, Transition from water to land in decapod crustaceans, *Am. Zoologist* **8:** 355–392.

Boolootian, R. A., Giese, A. C., Formanfarmaian, A., and Tucker, J., 1959, Reproductive cycles of five West Coast species of crabs, *Physiol. Zool.* **32**: 213–220.

Burkenroad, M. D., 1947, Production of sound by the fiddler crab, *Uca pugilator* Bosc, with remarks on its nocturnal and mating behavior, *Ecology* **28**: 458–461.

Cott, H. B., 1929, Observations on the natural history of the racing crab *Ocypode ceratophthalma* from Beira, *Proc. Zool. Soc. Lond.* **4**: 755–756.

Cowles, R. P., 1908, Habits, reactions and associations in *Ocypode arenaria*, *Papers Tortuges Lab. Carnegie Inst. Wash.* **2**: 1–41.

Crane, J., 1941*a*, Crabs of the genus *Uca* from the west coast of Central America, *Zoologica* **26**: 145–207.

Crane, J., 1941*b*, On the growth and ecology of brachyuran crabs of the genus *Ocypode*, *Zoologica* **26**: 297–310.

Crane, J., 1943, Display, breeding and relationships of fiddler crabs (Brachyura, genus *Uca*) in the northeastern United States, *Zoologica* **28**: 217–223.

Crane, J., 1957, Basic patterns of display in fiddler crabs (Ocypodidae, genus *Uca*), *Zoologica* **42**: 68–82.

Crane, J., 1958, Aspects of social behvior in fiddler crabs, with special reference to *Uca maracoani* (Latreille), *Zoologica* **43**: 113–130.

Crane, J., 1967, Combat and its ritualization in fiddler crabs (Ocypodidae) with special reference to *Uca rapax* (Smith), *Zoologica* **52**: 49–76.

Dumortier, B., 1963, Morphology of sound emission apparatus in Arthropoda, *in* "Acoustic Behaviour of Animals" (R. G. Busnel, ed.) pp. 277–345, Elsevier Publishing Co., New York.

Fellows, D., 1966, Zonation and burrowing behavior of the ghost crabs *Ocypode ceratophthalmus* (Pallas) and *Ocypode laevis* Dana in Hawaii, Unpublished M.S. thesis, University of Hawaii.

Griffin, D. J. G., 1968, Social and maintenance behaviour in two Australian ocypodid crabs (Crustacea: Brachyura), *J. Zool., Lond.* **156**: 291–305.

Harrigan, J., 1965, A circadian rhythm of locomotion in the ghost crab *Ocypode albicans* Bosc, Unpublished M. S. thesis, Duke University.

Horch, K., 1971, An organ for hearing and vibration sense in the ghost crab *Ocypode. Z. vergl. Physiol.* **73**; 1–21.

Horch, K. W., and Salmon, M., 1969, Production, perception and reception of acoustic stimuli by semiterrestrial crabs (genus *Ocypode* and *Uca*, family Ocypodidae), *Forma et Functio* **1**: 1–25.

Hughes, D. A., 1966, Behavioural and ecological investigations of the crab *Ocypode ceratophthalmus* (Crustacea: Ocypodidae), *J. Zool. Lond.* **150:** 129–143.

Linsenmair, K. E., 1965, Optische Signalisierung der Kopulationshöhle bei der Reiterkrabbe *Ocypode saratan* Forsk. (Decapoda Brachyura Ocypodidae), *Naturwissenschaften* **52:** 256–257.

Linsenmair, K. E., 1967, Konstruktion und Signalfunktion der Sandpyramide der Reiterkrabbe *Ocypode saratan* Forsk. (Decapoda Brachyura Ocypodidae), *Z. Tierpsychol.* **24**: 403–456.

Magnus, D., 1960, Zur Okologie des Landeinsiedlers *Coenobita jousseaumei* Bouvier und der Krabbe *Ocypode aegyptiaca* Gerstacker am Roten Meer, *Verh. Deutsch. Zool. Ges. Bonn*, pp. 316–329.

Miller, D. C., and Vernberg, F. J., 1968, Some thermal requirements of fiddler crabs of the temperate and tropical zones and their influence on geographic distribution, *Am. Zoologist* **8**: 459–470.

Salmon, M., 1965, Waving display and sound production in *Uca pugilator* Bosc, with comparisons to *U. minax* and *U. pugnax*, *Zoologica* **50**: 123–150.

Salmon, M., 1967, Coastal distribution, display and sound production by Florida fiddler crabs (genus *Uca*), *Anim. Behav.* **15**: 449–459.

Salmon, M., 1971, Signal characteristics and acoustic detection by the fiddler crabs *Uca rapax* and *Uca pugilator. Physiol. Zoölogy*, in press.

Salmon, M., and Atsaides, S. P., 1968*a*, Visual and acoustical signalling during courtship by fiddler crabs (genus *Uca*), *Am. Zoologist* **8**: 623–639.

Salmon, M., and Atsaides, S. P., 1968*b*, Behavioral, morphological and ecological evidence for two new species of fiddler crabs from the Gulf Coast of the United States, *Trans. Wash. Acad. Sci.* 81: 275–290.

Salmon, M., and Atsaides, S. P., 1969, Sensitivity to substrate vibration in the fiddler crab, *Uca pugliator* Bosc, *Anim. Behav.* **17**: 68–76.

Salmon, M., and Stout, J. F., 1962, Sexual discrimination and sound production in *Uca pugilator* Bosc, *Zoologica* **47**: 15–21.

Schöne, H., 1968, Agonistic and sexual display in aquatic and semiterrestrial brachyuran crabs, *Am. Zoologist* **8**: 641–654.

Watkins, W. H., 1967, The harmonic interval: Fact or artifact in spectral analysis of pulse trains, *in* "Marine Bioacoustics" (W. N. Tavolga, ed.) Vol. 2, pp. 15–44, Pergamon Press, London.

Chapter 3

RITUALIZATION IN MARINE CRUSTACEA

Brian A. Hazlett

Department of Zoology
University of Michigan
Ann Arbor, Michigan

I. INTRODUCTION

A. The Concept of Ritualization

The concept of ritualization of behavior patterns in animals is generally attributed to Huxley (1923), although Huxley (1966) gives credit for this concept to Selous (1901), and Darwin (1872) certainly had ideas along these lines. Ritualization has been variously defined (see Huxley, 1966), but it generally refers to the evolutionary development of communicatory movements or postures from movements or postures which originally served other purposes. Communication between individuals is improved by the evolutionary development of less ambiguous behavior patterns. Although the visual channel has been examined most frequently, evolutionary changes in sonic, tactile, and chemical activities could also result from ritualization, i.e., the development of a signal.

Many types of movement or other activities of animals can effect changes in the behavior of conspecific individuals, but generally a movement is thought to be ritualized only when there is evidenec that this activity has undergone changes, due to selective pressures, which improve its communicatory effectiveness. By definition, a ritualized pattern must have been, at one time, less functional in communication and not a display pattern or other specialized communicatory activity. Tinbergen (1952) considered activities occurring in conflict situations important source material for ritualization. During ritualization, the activities become "emancipated" from their original causal basis and function as independent units in a new context (communication)

(Tinbergen, 1952). The general outline of how ritualization must take place has been rather well worked out (Blest, 1961; Daanje, 1950; Morris, 1956, 1957; Tinbergen, 1952). Initially, some response by a recipient animal is made to an unritualized activity. The information transmission becomes more efficient when this activity becomes less ambiguous (by one of the methods to be mentioned), and selection can further reduce communications failure by limitations of the responsiveness of recipient individuals. Thus where the message being transmitted is potentially completely unambiguous, such as species identification, the signal may become very ritualized, if the presence of closely related species provides the necessary selection pressure. Situations in which there is room for some ambiguity, such as agonistic interactions, may not lead to such limitation of variability of movements. Thus an animal can be a member of only one species but can be slightly, moderately, or extremely aggressive.

Certain changes in movements might be expected if these are to function as communicatory elements (Daanje, 1950; Morris, 1966). Exaggeration of some components of movements, "freezing" of movements, omission of components, loss of coordination, changes in speed of execution, and decreases in variability of movement could occur during the evolutionary development of displays from nondisplay movements. The last change listed, decreases in variability or the development of stereotypy ("uniformity of motor behavior," Braestrup, 1966), has been termed "typical intensity" (Morris, 1957) or "typical form" (Tinbergen, 1964). "Typical form" may be the preferable term since it carries fewer connotations.

B. Methods of Study

Whatever the specific changes involved, the process of ritualization brings about the development of displays ("sensu," Moynihan, 1960) or other signals (Smith, 1965) from nondisplay movements. Thus all displays are ritualized patterns, although some may be more ritualized than others.

Several lines of evidence may be gathered by ethologists to indicate that ritualization has occurred, i.e., displays have developed. Clearly, the most direct evidence, observation over evolutionary time of changes in movements and in communicatory effectiveness, is not possible for most animals. The most common evidence given is (1) simply the description of a noticeable (to the ethologist) pattern and some observations concerning the situations in which the pattern appears. This is, of course, the necessary prerequisite of any work on a pattern. Next, (2) models may be used to examine the communicatory effectiveness of the display in an artificial situation and to clarify the sensory modality that the animals are responding to; without this step, the factors of importance to the animals can only be guessed at by the ethologist. In addition, (3) sequence analysis may be used to measure the communicatory capabilities of the displays more accurately and to verify that communication does take place in natural interactions. Finally, (4) comparison of the display movement with the nondisplay movement(s) it presumably evolved from may be carried out. This is frequently done in a descriptive fashion along with the ethogram (step 1) of the species. The comparison should be focused on those features of the display and nondisplay activities which model presentation have shown to be of importance to the animals. Comparative studies (Lorenz, 1950) can also be extremely useful in suggesting the source of ritualized movements.

This paper will briefly look at some of the levels of ritualization analysis carried out for marine Crustacea in general and then consider some measurements of behavior patterns of some hermit crabs.

II. COMMUNICATION STUDIES IN MARINE CRUSTACEA

A great number of "display" positions and other activities have been described for species of marine Crustacea. This summary will refer more to just the types of ritualized patterns seen in different groups rather than attempt a complete review.

A. Visual Agonistic Stimuli

Perhaps the most common display-like postures seen in agonistic situations in Crustacea are either a rapid movement of the chelipeds outward, i.e., spreading the chelipeds apart laterally ("lateral merus display," Wright, 1968), or an upward–forward movement of the chelipeds to where the surface of the mani of the chelipeds are presented to the other animal ("Aufbaumreflex," Bethe, 1897; "chela forward display," Wright, 1968). The former movement is common not only in brachyuran crabs (Crane, 1966; Schöne, 1961, 1968; Schöne and Schöne, 1963; Wright, 1968) but in some lobsters (Dybern and Höisaeter, 1965), stomatopods (Dingle and Caldwell, 1969), and shrimp (Johnson, 1969). In *Hemiplax hirtipes* (Beer, 1959) and *H. latifrons* (Griffin, 1968) crabs push back and forth with laterally spread chelipeds if initial execution of the Aufbaumreflex does not decide the encounter. The lateral spread of the chelipeds is accompanied by lateral extension of the ambulatory legs in *Mictyris longicarpus* (Cameron, 1966) as well as some spider crabs (Hazlett, unpublished observations; Schöne, 1968).

The second type of cheliped movement, chela forward display, is extremely common in brachyuran crabs (Beer, 1959; Daumer *et al.*, 1963; Griffin, 1965, 1968; Hiatt, 1948; Knudsen, 1960; Schöne, 1968; Schöne and Eibl-Eibesfeldt, 1965; Schöne and Schöne, 1963; Wright, 1968). Cott (1929) reported the movement of the bright red chelipeds of *Sesarma meinerti* high above the horizontal in a defensive situation. For further examples of postures in brachyuran crabs, see Schöne (1968) and Wright (1968).

B. Agonistic–Sexual Behavior Patterns

Some movements of marine Crustacea appear to be analogous to bird song, in that the patterns function as territorial–aggressive signals to other males and are also believed to act as a stimulating sexual signal to females of the species. The best known example of such patterns is the waving movements ("winken") of fiddler crabs, *Uca* spp. These movements have been well described for a number of species (Altevogt, 1955, 1957; Crane, 1941, 1943, 1957; Salmon, 1965, 1967; Salmon and Atsaides, 1968) and will not be discussed further here (the variability measures of Salmon, 1967, will be considered later). Analogous and perhaps homologous to the waving of male fiddler crabs are the movements of crabs in the genera *Heloecious* and *Hemiplax* (Griffin, 1965). The chelipeds are raised to a position high above the horizontal as the body tilts backward, and then the body and chelipeds are moved down (rapidly in *Heloecius*). Schöne (1961) discusses the possible evolutionary relationship of similar waving movements in forms such as *Dotilla* to the more elaborate waving of species of *Uca*.

An interesting variation of signal-level activity has been reported for the crab *Ocypode saratan* (Linsenmair, 1965, 1967). The male builds a pyramid of sand, near his burrow, which apparently acts as a visual signal showing the female the location of the male and the special spiral burrow. In addition, the pyramid acts as a territorial marker and stimulates other males to build new pyramids. Linsenmair (1967) discusses the possible evolution of this signal-level sand manipulation from the widespread sand-carrying movements of shore crabs.

C. Cleaning Symbiosis

An example of behavior associated with nonaggressive, nonsexual, non-predator–prey communication can be seen in the cleaning Crustacea such as shrimp of the genus *Periclimenes* (Limbaugh *et al.*, 1961). Many of these shrimp execute very noticeable lashing movements of their long antennae and their whole body rhythmically sways back and forth and side to side apparently as a signal to certain fish that "cleaning will be done here." This swaying and antennal movement are very probably used in communication and probably have been modified evolutionarily to serve this function.

D. Tactile Stimuli

Tactile stimuli used in communicatory situations may be divided into two categories—vibratory (equivalent to acoustical or sonic in animals with specialized hearing organs) and nonvibratory. Of the nonvibratory tactile stimuli, the clearest example of a signal-level effect is in the shell-fighting behavior of hermit crabs (Hazlett, 1966*a*,*b*, 1967). Here the tapping of the gastropod shell of an attacking crab against the shell of a defending crab may be followed by the defending crab leaving the shell. The tapping pattern is characteristic for each species, and in most species there is no direct physical contact between the bodies of the crabs. It was suggested (Hazlett, 1966*a*) that this tapping evolved from movements previously associated with the sudden application of physical force by an attacking crab.

In addition to the territorial waving movements previously mentioned, individuals of *Uca rapax* interact agonistically with other types of movements (Crane, 1967). While a few of the encounters Crane observed ended with the clear use of force, the majority of interactions were decided by apparently ritualized postures and pushing contests. In particular, major chelipeds of interacting crabs could become loosely interlocked and a series of light tapping of cheliped parts follow. The communication seemed to be largely by tactile stimuli acting at a signal level. If these ritualized tappings did not terminate an encounter, one crab was physically thrown back by the cheliped movement of the other.

A number of Crustacea produce sounds which could serve as communicatory elements. Guinot-Dumortier and Dumortier (1960) have summarized many of the examples of sound production in brachyurans. Additional examples of sound production in brachyuran crabs have been provided by von Hagen (1968) and Salmon (1965, 1967) and Salmon and Atsaides (1968). The production of stridulatory sounds in agonistic situations has been reported in several species of land hermit crab, *Coenobita* (Hazlett, 1966*c*; Magnus, 1960). The rasping sounds of various panulirid lobsters (Dijkgraaf, 1955; Hazlett and Winn, 1962*a*,*b*; Lindberg, 1955; Moulten, 1957) have been described and communicative functions suggested for some. The nonstridulatory sound production of *Homarus* (Fish, 1966) may prove to be of special interest from a physiological viewpoint.

Communication prior to copulation would seem to be of extreme importance, since lack of species identification could result in wastage of genetic material. Tactile signals could be exchanged during the precopulatory cheliped pushing of *Pachygrapsus* (Bovbjerg, 1960) or the precopulatory leg tapping of male *Heloecius cordiformis* (Griffin, 1968). Certainly in hermit crabs (Hazlett, 1966*a*, 1968*a,b*) the species-characteristic rocking and tapping of the female by the male are acting at the signal level, as is the palpating of the male's mouthparts by the female just prior to copulation.

E. Chemical Stimuli

While pheromone production has not been considered in most discussions of ritualization, it is clear that information transmission along the chemical channel can be improved by changes in both chemicals produced and those responded to. Recent experiments on the hermit crab *Pagurus bernhardus* (Hazlett, 1970*a*) indicated the presence of a male-stimulating chemical on females. But the only clear evidence for a diffusible pheromone in Crustacea has been Ryan's (1966) study of *Portunas sanguinolentus.* He showed that the male is sexually stimulated by a water-soluble substance produced by premolt females.

III. MODEL PRESENTATIONS AND SEQUENCE ANALYSIS

Models have been used in a few cases to show that animals were reacting to certain stimuli associated with a given display. Hazlett and Winn (1962*a*) presented closed and open claws of snapping shrimp (*Alpheus* and *Synalpheus*) to individual snapping shrimp and observed many more agonistic reactions to the open claw. Simulation of the water jet produced by an alpheid's "snap" also elicited an increase in agonistic behavior (Hazlett and Winn. 1962*a*), although playbacks of recorded snaps to shrimp, speaker in contact with substrate, had no observable effect on their behavior (Hazlett, unpublished data). Salmon and Stout (1962) used clay models with attached chelipeds to demonstrate sexual recognition by the presence or absence of the large cheliped in *Uca pugilator*. The releasing effect of female movement in *Uca tangeri* was tested by von Hagen (1961) with a moving model. Salmon (1965) used sound playbacks to demonstrate the effectiveness of *U. pugilator* sounds in increasing sound production by conspecific males. Salmon and Atsaides (1969) used playbacks to demonstrate responsiveness of female *U. pugilator* to sounds such as those produced by males.

Model presentations to hermit crabs will be considered in the following section.

The use of some statistical technique to analyze sequences of behavior would seem to be a logical step in the study of displays, since the actual occurrence and magnitude of communication (the function of displays) can thus be demonstrated. Until this step is taken, the supposed communicatory action of a behavior pattern can only be tentatively assumed (see Dingle, this volume), and any discussion of ritualization would be premature until this assumption is verified. The use of models can demonstrate the use by the animals of a given sensory modality, but such experiments cannot prove or measure the existence or extent of communication in conspecific interactions. Unfortunately, sequence analysis requires a considerable amount of data; thus only two groups of Crustacea have been examined in this regard.

Hazlett and Bossert (1965, 1966) measured the magnitude of change in social behavior following the execution of agonistic display movements by consepcific crabs in nine species of hermit crab. Dingle (1969) carried out a similar study on the stomatopod *Gonodactylus bredini.* In both types of Crustacea, the assumption of a given posture by one member of an interacting pair was followed by a number of behavior patterns in the other animal. Analysis of the distributions of following acts showed highly significant changes in the probability distributions of acts dependent upon the preceding act. That is, execution of displays did change the behavior of recipient individuals. Historical context (Smith, 1965) effects were also demonstrated—what an animal did in response to a given pattern was dependent upon earlier acts. Interestingly, the magnitude of change, as measured by information formulae analysis, was of the same order of magnitude for hermit crabs and stomatopods (average 0.41 bits per display transferred for the eight species of hermit crab, 0.78 bits per display for the stomatopod). This is somewhat below the measure given for rhesus monkey (1.96 bits) by Altmann (1965); however, since Altmann lumped inter- and intraindividual sequences (p. 496), it is difficult to equate his study with the crustacean analyses (see Dingle, this volume, for further discussion). Dingle (1969) also presents data which show an increase in the information transmitted by displays as dominance–subordinance relationship are established ($H_t = 0.63$ bits per display during the first 10 min of interaction, $H_t = 1.03$ bits per display during the second 10 min).

IV. RITUALIZATION IN HERMIT CRABS

Hermit crabs (superfamily Paguridea) communicate by use of visual displays (Hazlett, 1966*a*,*b*; Hazlett and Bossert, 1965; Reese, 1962, 1964), although other activities and modalities are also used in social communication (Hazlett, 1968*c;* 1970*a*). Over 60 species have been observed, and while some show considerable elaborations (Hazlett, 1968*d*, 1969) almost all have three basic agonistic postures (Fig. 1): a lateral movement of one or more ambulatory legs (ambulatory raise), a low-level agonistic move-

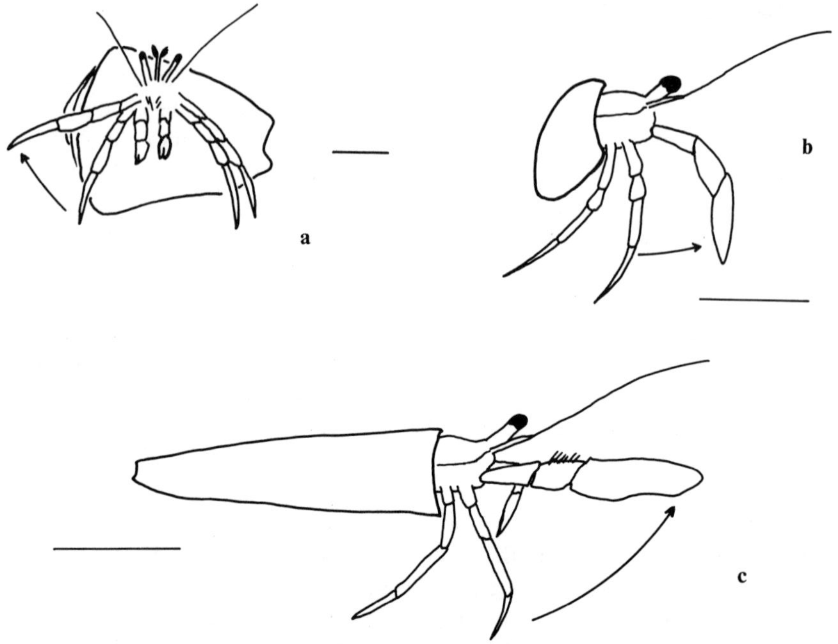

Fig. 1. Outline of agonistic display postures of hermit crabs. (**a**) Single ambulatory raise by *Paguristes puncticeps.* (**b**) Cheliped presentation by *Anapagurus laevis.* (**c**) Cheliped extension by *Pylopagurus gibbosimanus.* All scale lines equal 1 cm.

ment of one or both chelipeds forward to a position where the mani are approximately perpendicular to the substrate (cheliped presentation), and a higher- intensity agonistic movement of one or both chelipeds to a position where the mani are a bit below the horizontal (cheliped extension). It was suggested (Hazlett, 1966*a*) that the ambulatory display has been evolved from walking–climbing movements and the cheliped displays from feeding movements. In some species, including *Clibanarius vittatus*, sequence analysis (Hazlett and Bossert, 1966) has shown that these postures (ambulatory raise, cheliped presentation, cheliped extension) are effective in communication. In addition, the presentation of models, with the limbs in various positions, has shown these postures to be effective visual stimuli (Hazlett, 1966*a*, 1968*d*). Experiments with a model with a movable cheliped indicated that position is more important than movement at least in one species, *Calcinus tibicen* (Hazlett, 1966*a*).

The following measurements and experiments were carried out to examine ritualized movements in two species of diogenid hermit crab—*Clibanarius vittatus* and *Petrochirus diogenes*. Both are shallow-water crabs found in southern Florida and elsewhere in the Caribbean. The social behavior of

C. vittatus has been described elsewhere (Hazlett, 1966*a*), and several aspects of social interaction have been dealt with experimentally (Hazlett, 1968*c*, 1970*b*).

A. Agonistic Movements in *Petrochirus*

Petrochirus diogenes is a very large hermit crab, often occupying shells of *Strombus gigas* (Provenzano, 1959). Specimens were initially studied in the laboratory at the Institute of Marine Science in Miami and in the field around Key Biscayne, Florida, during portions of 1964 and 1965. These are very active hermit crabs, moving about rapidly with the gastropod shell held well off the substrate. The ambulatory legs are in almost continual motion either in locomotion or in feeding; in the latter, the ambulatory legs are moved out to the side a little and then pulled in, bringing sand toward the mouth region of the crab. When a crab is actively feeding, a steady stream of sand grains and other particles is pushed forward in front of it by the movements of the feeding appendages.

While it will be demonstrated that *Petrochirus* is not an especially social species, specimens will show visual social facilitation (Zajonc, 1965) of feeding movements. When one individual sees a conspecific feeding, it shows feeding movements itself. If a *Petrochirus* is chemically stimulated by introducting a soluion of fish juice, it waves its ambulatory legs back and forth and increases its locomotory activity, reactions similar to the patterns shown by *Clibanarius vittatus* (Hazlett, 1968*e*). To test for social facilitation of feeding activity, one all-glass 2 gal aquarium was placed inside a 10 gal aquarium, water levels were adjusted to just below the level of the smaller tank, and a *Petrochirus* was placed in each aquarium. After a 15 min adjustment period, the bouts of leg waving and rate of locomotory activity of the "observer" crab were counted for 1 min periods before and after adding filtered fish juice to the smaller aquarium containing the "reacting crab." The glass tube used to add fish juice was raised in the same manner at the beginning of both minutes of observation, but fish juice was introduced only at the start of the second minute. The results for 12 pairs are shown in Table I.

The sight (or possibly sound) of an active "reacting" conspecific clearly influences the feeding activity of an "observing" crab. During the minute the reacting crab was more active, the observing crab moved about the aquarium more and waved its ambulatory legs more.

When two individuals of *Petrochirus diogenes* move close to one another, movements of the massive chelipeds and ambulatory legs often occur. However, it appeared that (1) the limbs were not moved to any certain position but rather that extent of movement was highly variable, and (2) very frequently the limbs physically struck the other crab with considerable force.

Table I. Changes in Feeding Behavior of a *Petrochirus diogenes* Which Observed a Conspecific Crab Stimulated by Fish Juice[a]

	Locomotory activity, lines crossed	Bouts of ambulatory waving
Before	4	7
After	14	25
	$\chi_c^2 = 4.5$	$\chi_c^2 = 9.0$
	$p < 0.05$	$p < 0.005$

[a] Summed data for 12 pairs.

Execution of a cheliped or ambulatory leg movement by one crab was usually followed by a similar movement and/or retreat by the other crab. "Retreat" was frequently the direct result of being hit by a cheliped. The cheliped of one crab often hit the other crab either on the way up or as the appendages were rapidly brought down. The ambulatory leg movements were often directed forward as well as up and struck the other crab in the eye region. As in other species of hermit crab, individuals often retreated just from the approach of a conspecific. It was difficult to maintain two or more individuals of *Petrochirus* in the same tank because of the physical damage done by conspecifics to each other, During a 2 week period, one large individual killed all ten conspecifics in the 10 gal tank with it, although considerable food was provided for all the crabs. The massive chelipeds are very strong, and even large crabs are not immune to having their chelipeds crushed (and then autotomized) and eaten by slightly larger conspecifics. Frequently, larger crabs were observed to tear, one by one, all the limbs from a smaller one and then pull the body out of the gastropod shell and consume it. *Petrochirus* does show rather typical (albeit very forceful) shell-fighting behavior, and exchanges of shells following a series of rocking and rapping movements (see Hazlett, 1966*a*) do occur. But, also, large crabs have been observed to execute the raps associated with shell fighting against a much smaller crab (in a shell the larger could not possibly get into), and when the smaller crab crawls out of its shell the larger crab seizes it and consumes it, pushing the gastropod shell to one side. During rapping the two shells are brought together with considerable force, and often the chelipeds are moved down forcefully on the chelipeds of the defending crab. Many of the raps executed in *Petrochirus* shell fights involve movement of both the attacker's and the defender's shell (by the attacker) as the two are brought together. In regular "diogenid" rapping only the attacking crab's shell is moved, while only the defender's shell is moved in "pagurid" rapping–shaking (Hazlett, 1966*a*).

While these acts of physical damage may be laboratory induced in some way, similar population densities of other species of hermit crab may survive for many weeks in the laboratory without any such incidents of conspecific damage. It should be mentioned that in the field, *Petrochirus* is very widely dispersed. It is rare to find more than one individual in the same square meter of bottom. This type of spacing out is not common among other shallow-water hermits, many of which form large aggregations (Ball, 1968; Hazlett, 1966*a*). It is very common to find five to ten specimens of *Clibanarius vittatus* feeding on a mud flat in the same square meter. In addition, some species, such as *Paguristes puncticeps*, are just as widely distributed in the field as *Petrochirus*, and yet when groups were just as "crowded" in the laboratory no intraspecific damage occurred.

To obtain some measure of the difference in conspecific damage done in the two species (*Petrochirus* and *C. vittatus*), animals were maintained in the laboratory (at the University of Michigan, specimens flown in from Florida) under two social conditions—isolated and nonisolated (three conspecifics per container). One, two, and ten gallon aquaria were used, but no effect of container size was seen in the limited number of replications run. In all cases, the three animals in any nonisolated situation were closely matched for size, but despite efforts to match sizes of the crabs of the two species the specimens of *Petrochirus* averaged about 10 mm (40%) longer in cephalothorax length than those of *Clibanarius vittatus*. Each aquarium was supplied with a undergravel filtration–aeration system, and each crab was given a standard-size piece of fish every 3 days. The aquaria were examined daily. Molts were noted and dead crabs examined. Preliminary results in days before one crab in a container was dead are given in Table II; additional replicates are currently in progress.

It is clear that the difference in survival in isolated and nonisolated conditions is different for the two species. The decrease in length of survival was much more in *Petrochirus* (65 percent) than in *Clibanarius* (24 percent). Many controls would be needed to verify that differences in social communication are causally related to the differences in survival. However, in most of the nonisolated crab deaths in *Petrochirus* the remains clearly indicated considerable physical damage by conspecifics. And in only one case did a *Petrochirus* survive a molt. Conspecific attack was observed to be the cause of death in three of the nine replicates. In *Clibanarius* no evidence of physical

Table II. Length of Survival (in Days) in Laboratory of Isolated and Nonisolated (Three Crabs per Container) Hermit Crabs[a]

	Isolated	Nonisolated
Clibanarius vittatus	127 (103–201)	97 (89–106)
Petrochirus diogenes	80 (36–122)	28 (7–58)

[a] Average, with range given in parentheses.

Table III. Reactions Shown by Specimens of *Petrochirus diogenes* to the Presentation of Visual Models[a]

Models	Reactions to model: Retreated	Executed aggressive movement	No observable reaction	Number of animals tested
Double cheliped extension	25	10	15	
Control, neutral model	27	7	16	50
Double ambulatory raise	20	7	18	
Control, neutral model	20	4	21	45

[a] Individuals were tested once with each model.

damage by conspecifics was found, and a number of molts occurred without crabs being attacked.

To initially test for the presence of display postures, models were constructed (Hazlett, 1966*a*) from dried exoskeletons of specimens of *Petrochirus*. The limbs were fixed in the desired positions as the exoskeletons dried and were held in place by waterproof glue. The models were attached to glass rods and presented (Hazlett, 1968*b*) to specimens of *Petrochirus* in individual containers. The models, which were all the same size, were (1) a control model with the limbs in a neutral position (chelipeds under mouth area, ambulatory legs in normal walking position), (2) a cheliped extension model (ambulatory legs normal, both chelipeds forward, horizontal to substrate), and (3) an ambulatory raise model (chelipeds under the mouth area, both first ambulatory legs out to the side, horizontal to the substrate). The responses of the individuals tested are shown in Table III. The responses to the test models were not different from the responses to the control models. Individuals of *Petrochirus diogenes* do not appear to react to limb postures which elicit agonistic behavior in other species of hermit crab. The results of a similar test with *Clibanarius cubensis*, a species very closely related to *C. vittatus*, are presented in Table IV for comparison.

Table IV. Responses of 50 Individuals of *Clibanarius cubensis* to Models with Different Limb Positions (from Hazlett, 1966*a*)

Models	Reactions to model: Retreat	Execute aggressive movement	No observable response	Chi-square response–no response
Cheliped extension	39	1	10	9.90, $p < 0.01$
Control	25	0	25	
Ambulatory raise	38	1	11	16.42, $p < 0.01$
Control	18	1	31	

B. Film Analysis

One of the changes that can occur during ritualization is the development of stereotypy (Morris, 1957, 1966). To obtain some measure of this and other characteristics, motion picture films of locomotory movements and agonistic interactions between conspecific crabs were taken of *Clibanarius vittatus* and *Petrochirus diogenes*.

The crabs were filmed in Ann Arbor, Michigan, following shipment from Big Pine Key, Florida. Approximately 100 specimens of *C. vittatus* and 30 specimens of *P. diogenes* were used. Films were taken with a Bolex H-16 equipped with a 75 lens and extension tubes while the crabs interacted in a 10 gal aquarium with sand and gravel on the bottom. The aquarium was near a window and no special lighting was used. Nine hundred feet of film at 16 frames per second was shot of *Petrochirus* and 1000 ft of *Clibanarius*.

The films were analyzed frame by frame for duration of movement and changes in position using a Bell & Howell time and motion study projector. Position measurements were made from enlarged drawings traced from the projection of individual frames. Final position measurements for a movement used the frame in which the limb in question had moved to the greatest extent in the direction involved. Two classes of movements were measured, locomotory movements and agonistic movements. The latter were taken as any movements of limbs executed as two crabs came near one another that were not clearly locomotory or feeding movements (displacement feeding movements were not counted as agonistic for purposes of these measurements). No classification of movements into the display categories previously described (Hazlett, 1966*a*,*b*) was done, since this would bias the results greatly. That is, some of the movements of the chelipeds which occurred in agonistic situations were not clear "cheliped presentation" or "cheliped extension" displays. Nevertheless, all movements were included in the analysis, thus making the data messier than if cleaned up by exclusion of patterns that "did not fit."

Only those sequences of movements in which the crab was either directly facing the camera (ambulatory leg movements) or was at right angles to the camera (cheliped movements) were used for measurements. Since some movements, particularly in *Petrochirus*, contain motion in several directions, some postural information (felt to be of less importance to the crabs) was essentially ignored in the following analysis. I decided that it was impractical to try to estimate positions from sequences in which the animal was at intermediate positions relative to the camera. These restrictions meant not using many sequences with either an intermediate angle or a wrong angle (sequences correct for ambulatory leg movements were wrong for cheliped movements and vice versa). Individuals of *Clibanarius* and *Petrochirus* readily walk forward, backward, or to either side. However, in the analysis of locomotory movements, only movements to one side or the other and only movements of legs on the side toward which the crab was locomoting were used.

The position measurements were taken as outlined in Fig. 2. Ambulatory leg position was taken as the angle formed between a line connecting the middle of the dactyl of the involved leg and a point midway between both eyestalks and third maxillipeds and a line in the sagittal plane of the animal. Cheliped position was taken as the angle formed between a line connecting the middle of the manus of the involved limb and a point just below the ipsilateral eyestalk and a line parallel to the animal's anterior–posterior axis, running through the point just below the ipsilateral eyestalk.

The distributions of final limb positions for the two classes of movements for the two species are shown in Figs. 3 and 4. The averages, standard deviations, and coefficients of variability of the final position and duration of movements of the different movements are presented in Table V. The differences in the variances of the movements were tested using F ratios for the final position and duration of movements, and these results are presented in Table VI. Relationship of mean and SD is shown in Fig. 5.

It is clear from these measurements that, for *C. vittatus*, movements that were shown to be effective in communication are more stereotyped (i.e., less variable) than a nondisplay movement such as locomotion (the class of movements suggested as the evolutionary source for the ritualized leg movement; Hazlett, 1966*a*). For the cheliped movements, the differences in variance of both position and duration of movements were very significant. In the case of ambulatory leg movements, the difference in position variance was barely statistically significant, while differences in duration of movement were not. It is perhaps important that the movement with the highest degree of stereotypy, the cheliped extension ($\bar{V} = 4.0$), is also the movement which

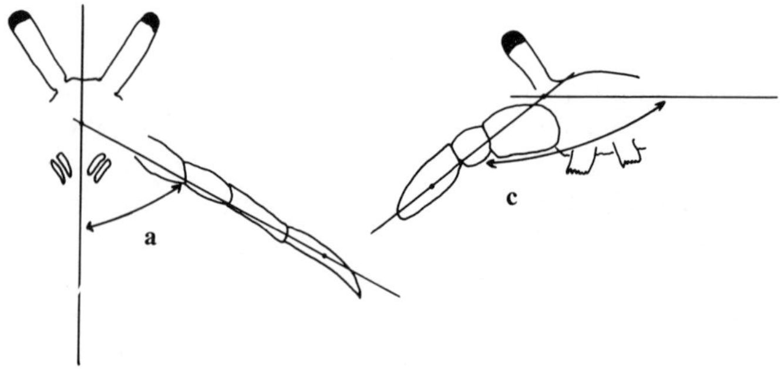

Fig. 2. Diagram of how limb position measurements were taken for ambulatory leg movements **(a)** and cheliped movements **(c)**.

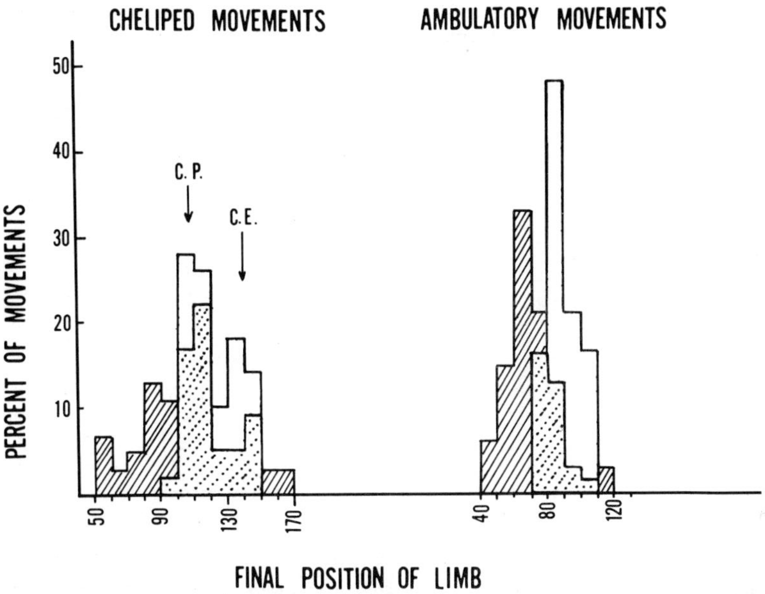

Fig. 3. Distribution of limb movement positions for *Clibanarius vittatus;* cheliped movements on the left, ambulatory leg movements on the right. Walking movements are represented by lined area, agonistic display movements by clear area, overlap areas by dotted area. Plotted as percent of total movements of that class per 10° segment.

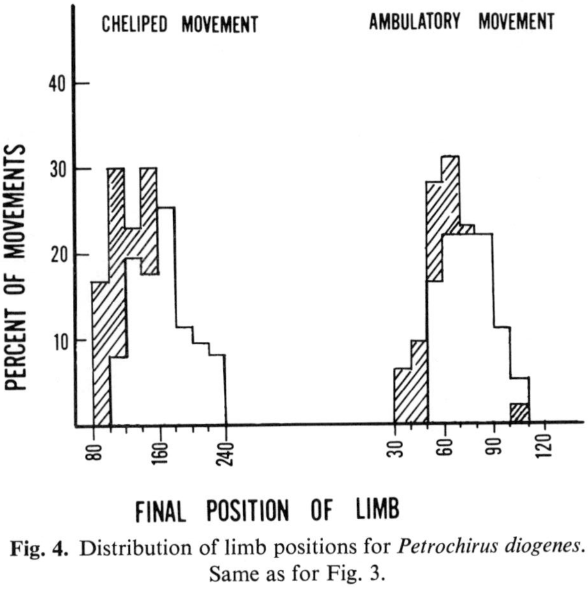

Fig. 4. Distribution of limb positions for *Petrochirus diogenes.* Same as for Fig. 3.

Table V. Walking and Agonistic Limb Movements, Final Position, and Duration of Movements in Frames of Film

		Final position (degrees)			Duration	
	N	Mean	SD	Coefficient of variability	Mean	SD
Clibanarius vittatus						
Cheliped—walking	44	105	25.4	24.1	7.7	3.1
Cheliped—all agonistic	48	120	14.1	11.8	—	—
Cheliped presentation	34	112	7.7	6.9	3.5	1.4
Cheliped extension	14	138	5.5	4.0	4.5	1.6
Ambulatory—walking	65	70	14.0	20.1	5.1	2.3
Ambulatory—agonistic	19	88	10.1	11.6	5.3	1.9
Petrochirus diogenes						
Cheliped—walking	23	123	22.9	18.6	15.9	7.9
Cheliped—agonistic	61	164	33.0	20.1	5.2	2.8
Ambulatory—walking	32	63	14.4	23.0	12.3	6.5
Ambulatory—agonistic	18	75	14.2	19.1	9.4	8.8

in this species effects the greatest change in the behavior of a recipient crab (Hazlett and Bossert, 1966). For *Petrochirus diogenes*, the only significant differences in variance were in the final cheliped position (where the agonistic movement was *more* variable) and duration, where the walking movement was much more variable. It would appear that agonistic movements are not ritualized in *Petrochirus*. Certainly, "typical form" is not readily seen in their agonistic movements.

It is important to keep in mind that the variance in position being examined in this study is the factor indicated to be of importance by other tests. It

Table VI. Comparison of Variances of Locomotory and Agonistic Movements

	Final position		Duration	
	F ratio	p	F ratio	p
Clibanarius vittatus				
Cheliped				
Walking—all agonistic	3.2	<0.005	—	—
Walking—presentation	10.7	<0.005	4.6	<0.005
Walking—extension	21.3	<0.005	3.6	<0.005
Ambulatory				
Walking—agonistic	1.9	$=0.05$	1.5	>0.10
Petrochirus diogenes				
Cheliped				
Walking—agonistic	1.0	>0.10	7.9	<0.005
Ambulatory				
Walking—agonistic	2.1	<0.05	1.8	>0.05

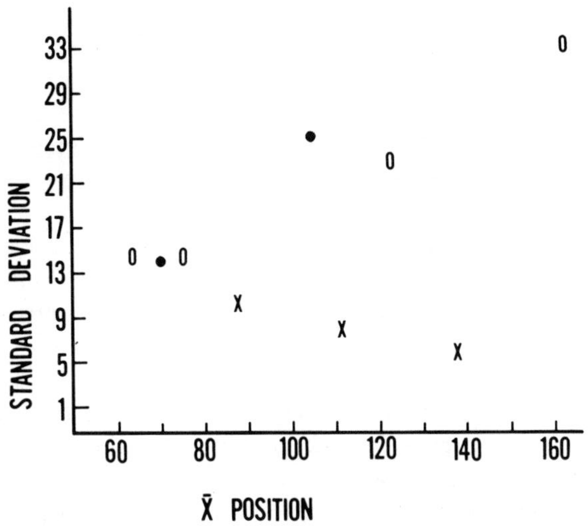

Fig. 5. Relationship of mean position of limb and standard deviation for positions for hermit crab behaviors. Displays of *Clibanarius vittatus* (X), nondisplays of *C. vittatus* (●), and movements of *Petrochirus diogenes* (O).

would be of limited interest to measure variability in a parameter of behavior not known to be of importance to the species studied.

The agonistic movements of *Clibanarius vittatus* can now be described more precisely. An ambulatory raise is a movement of one or more ambulatory legs outward to a position 88 ± 10° lateral to the perpendicular, outward movement taking 0.31 ± 0.11 sec to execute. The cheliped presentation is a movement of one or both chelipeds forward to a position where the manus is 22 ± 7° beyond a line perpendicular to the substrate, forward movement taking 0.21 ± 0.10 sec. A cheliped extension is a movement of one or both chelipeds forward and upward to a position 48 ± 5° beyond the perpendicular, taking 0.27 ± 0.11 sec to execute.

In addition to data on variability of movements, these films also provide information on other differences between display movements and nondisplay movements in *C. vittatus*. As noted by Daanje (1950), exaggeration of one component of a movement may occur during ritualization. The outward movement of the ambulatory legs during the execution of an ambulatory raise display is greater in both final position (88° from the perpendicular compared to 70° average for walking movements) and degrees of arc traversed during movement (average 49.5° for agonistic ambulatory movements, average 28.6° for walking movements), although the range of nondisplay movements is greater than that of display movements (see Fig. 3). Similarly, the average final position of the chelipeds is greater for display movements than for nondisplay movements, although the range for nondisplay movements is greater.

Comparison by *t*-tests of the means of final positions of limbs during walking and agonistic movements of *C. vittatus* showed that the average positions during agonistic movements were different (greater). The *t*-test value for ambulatory movements equals 5.3, $p < 0.001$, and the *t*-test value for cheliped movements equals 3.5, $p < 0.001$.

Another characteristic of ritualized movements mentioned by Morris (1966) is the "freezing" of components of the movements. In the movements concerned with walking in both species and in the majority of agonistic movement of *Petrochirus*, the number of frames of film showing the limb in the final position was very small. In movement of the chelipeds of *Petrochirus* in which the limb was moved beyond 170° from the anterior–posterior axis, the chelipeds were never held in position, but always started to descend in the frame just after that of the highest position. In contrast, the ambulatory legs of *C. vittatus* were held in the raised position between 0.30 and 2.22 sec (5–37 frames), while the chelipeds were held in the raised position between 0.24 and 1.08 sec (4–18 frames) for the cheliped extension display and between 0.24 and 6.72 sec (4—112 frames) for the cheliped presentation display.

Another characteristic of ritualization mentioned by Morris (1966) is changes in the speed of execution of movements. For both species studied, the agonistic movements of the chelipeds were executed more rapidly than the walking movements of the chelipeds. There was no difference for the ambulatory movements of *C. vittatus.*

It appears from the results of model presentations, sequence analysis, and movement variability measurements that *Clibanarius vittatus* communicates at a display level, while the available evidence does not indicate that *Petrochireus diognes* does. Among the 60 species of hermit crab I have observed, the latter type of nonritualized communication seems to predominate in just a few species. These are *Dardanus venosus* (Miami and the Florida Keys) [actually *D. venosus* is probably two species (Provenzano, personal communication), but both very regularly use physical force in social interactions], *D. arrosor* and *D. callidus* (Naples, Italy), and two undescribed *Dardanus* species from the Pacific coast of Panama. *Dardanus* and *Petrochirus* are in a separate subfamily, Dardaninae, of the family Diogenidae. All other species of hermit carb in the Diogenidae (those in the subfamily Diogeninae) that I have observed (*Clibanarius cubensis, C. tricolor, C. antillensis, C. erythropus, C. panamensis, C. anomalus, Calcinus tibicen, Cal. verrilli, Cal. ornatus, Diogenes pugilator, Paguristes grayi, P. cadenati, P. puncticeps, P. oculatus, P. morrei, P. tortugae, P. sayi, P. depressus, P. tri-*

angulatus, *P. hummi*, *P. spinnipes*, and *P. anomalus* plus several undescribed species) as well as about 30 species of Paguridae communicate primarily by means of visual displays. In all species, except those in the Dardaninae, it is easy to recognize relatively stereotyped movements which occur in agonistic situations. In all these species, except the Dardaninae, intraspecific damage during fighting is rare even under rather crowded conditions.

In contrast, individuals of *Petrochirus diogenes*, *Dardanus venosus*, *D. arrosor* and *D. callidus* very frequently do physical damage to conspecifics under laboratory conditions. [An exception in the Dardaninae seems to be *Dardanus insignis* (Hazlett, 1966*b*), which appears to have recognizable postures and which did not show high levels of conspecific damage in the laboratory.] Even when there were only two crabs per 10 gal tank, not one specimen of *D. arrosor* survived a molt during 3 months of observation in Naples. And many were eaten by larger conspecifics during intermolt periods. It is noteworthy that the ambulatory leg movements used by specimens of *D. arrosor* to stimulate sea anemones, *Calliactis parasitica*, to release their hold on the substrate (Brock, 1927; Ross and Sutton. 1961, 1968) appear to be much more stereotyped than intraspecific agonistic movements.

C. Multilevel "Typical Form" in *Pagurus longicarpus*

In his paper on the characteristics of ritualized movements, Morris (1957) proposed the term "typical intensity" to describe the relationship between stimulus strength and intensity of movement in the case of displays. Figure 6 is an adaptation of Morris's plot to show the relationships for a behavior pattern showing "typical form." This relationship can also be plotted (as in the present study) as a distribution of extent (intensities) of movements of animals exposed to a distribution of varying stimulus intensities (Fig. 7).

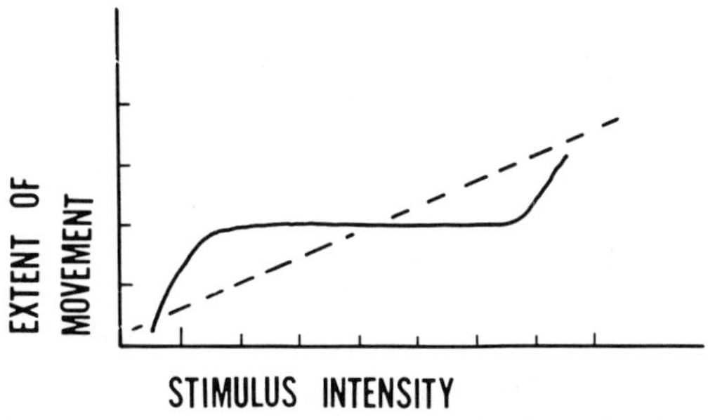

Fig. 6. Relationship of extent of movement to stimulus intensity for a behavior pattern showing "typical form" (solid line) and one showing variable form (dotted line).

The latter can be assumed to be the case in this study for the crabs as they wander about an aquarium coming within social contact distance of various sizes of conspecific individuals.

For agonistic interactions in hermit crabs, the distribution of stimulus intensities may not be really normally distributed, since there is a cut-off of sizes (too large, too small) that a given crab (*Clibanarius vittatus*) will react to (Hazlett, 1968*c*).

Morris (1957) concentrated on the case where there is just one type of movement associated with a given stimulus situation—such as the ambulatory raise display of *C. vittatus*. However, the relationships of intensity of stimulus, extent of movement, and frequency of various movements can be more complex than shown in Figs. 6 and 7. [The cheliped presentation–cheliped extension bimodal distribution (a two-step intensity–extent curve) for agonistic movements of the chelipeds in *C. vittatus* is one example.] In the pagurid hermit crab *Pagurus longicarpus*, a trimodal distribution is apparent (Figs. 8 and 9). There are a cheliped presentation movement, a cheliped extension, and a "forward cheliped movement" which is intermediate in position between the presentation and extension. According to the method of description outlined in the film analysis section, the three movements would bring the major cheliped to positions about 95°, 120°, and 155° from the anterior–posterior axis. *P. longicarpus* is a common shallow-water hermit which I observed both in Massachusetts and Florida. After initial ethological observations, a series of mirror presentations to specimens collected on the west coast of Florida was used to test the relationship of stimulus intensity to extent of movement.

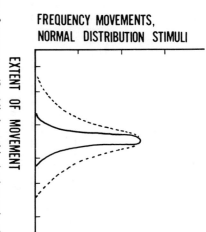

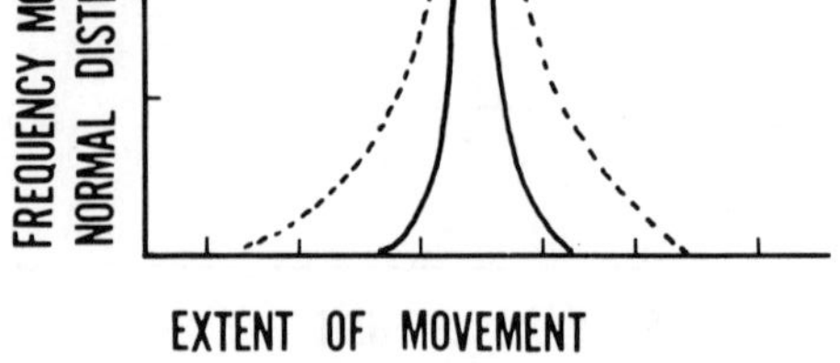

Fig. 7. Distribution of movement (final limb position) when animal is exposed to normal distribution of stimulus intensities for a behavior pattern showing "typical form" (solid line) and variable form (dotted line).

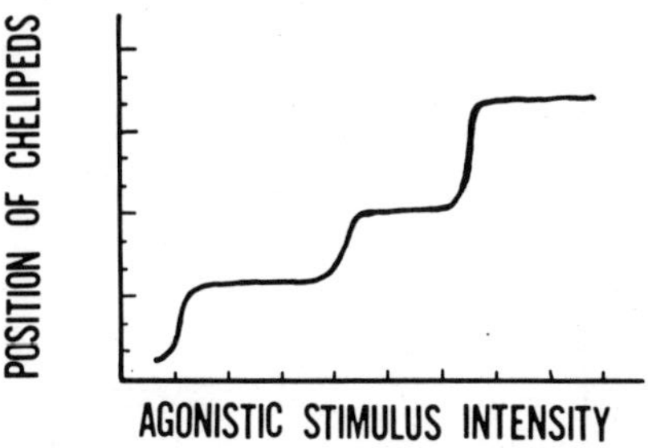

Fig. 8. Relationship of stimulus intensity and extent of movement of appendage for a set of three related behavior patterns, each of which shows "typical form." Agonistic cheliped movements of *Pagurus longicarpus*.

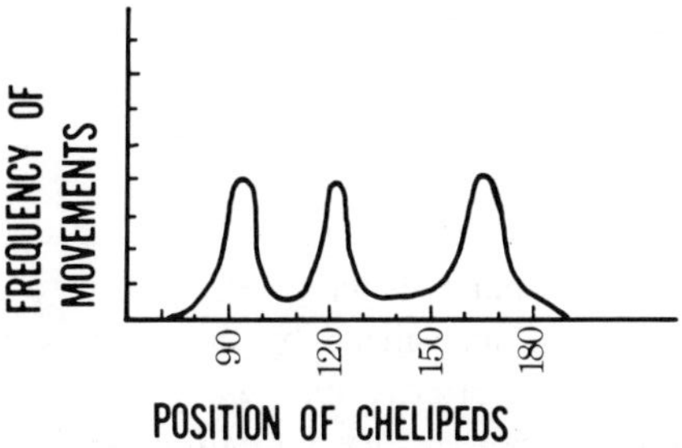

Fig. 9. Relationship of position of chelipeds and frequency of movement to that position for *Pagurus longicarpus* exposed to normal distribution of agonistic stimuli.

Placing a small mirror in front of crabs elicits a spectrum of agonistic reactions (Hazlett, 1966*a*). In this experiment, each crab was placed in an individual culture dish with sand on the bottom. A 3 by 4 inch mirror was placed in three positions in front of the crab after it was up in its shell (i.e., not withdrawn): two cephalothorax lengths (c.l.), 1 c.l., and $\frac{1}{2}$ c.l. in front of the crab. The sequence of positions for each of the 36 crabs tested was determined from a random number table. Of the 36 animals, 14 showed no response to the mirror in any position. For the 22 crabs that responded, the distribution of cheliped movement for the three stimulus situations is shown in Fig. 10. Twenty of the 22 crabs showed an increase in extent of cheliped movement as the mirror distance was decreased. The data could be plotted in the manner of Figs. 3, 4, and 6 and the resulting distribution would be similar to that shown in Fig. 9. Such a trimodal distribution for all movements (of chelipeds) occurring in a social situation (agonistic) can be just as ritualized (relatively small variance for each movement) as a unimodal single-movement distribution.,

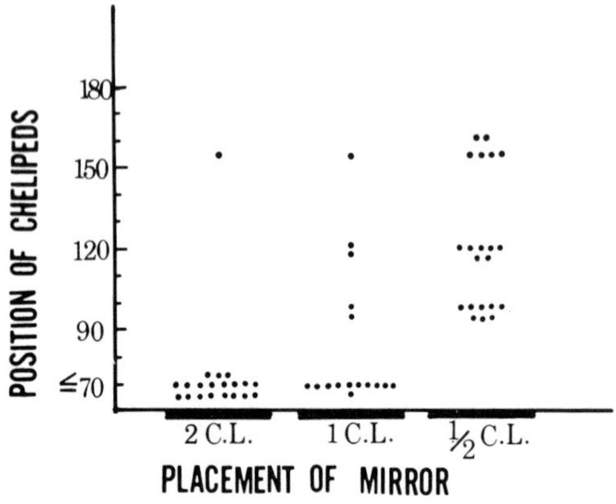

Fig. 10. Relationship of extent of cheliped movements by individuals of *Pagurus longicarpus* and proximity of mirror placed in front of crab.

V. DISCUSSION

A number of behavior patterns have been described in marine Crustacea which probably have been specialized by the process of natural selection, for communication. Such patterns may be effective through the visual, auditory–tactile, or chemical sensory modalities and can be important in agonistic, sexual, or interspecific symbiotic interactions. The process of improving communication betweeen individuals by the evolutionary development of less ambiguous behavior patterns is ritualization. Those individuals which do not reduce their fitness by directing sexual behavior at an incorrect species may be at a selective advantage. Animals which engage in frequent hard physical contact will be at a disadvantage compared to those which are able to gain the needed environmental resources without physical damage from conspecifics. In most situations, a loser in a ritualized fight has a much better chance to reproduce than a loser in an unritualized fight, particularly if the combatants have potential weapons such as very large chelipeds.

The study of ritualized behavior patterns involves an ethological description of the pattern, tests of several kinds to show that the pattern is effective in communication, and measurements and comparisons of the pattern with other patterns to suggest its origin. As indicated in Section II, the majority of work for marine Crustacea has been at the initial descriptive level. The number of cases in which the communicative effect of activities has been measured is small, and experimentation on communication has been largely confined to two groups—hermit crabs and fiddler crabs. The postures assumed by hermit crabs during agonistic interactions have been shown to be visually effective in communication, and the magnitude of communication has been measured for nine species. In the present paper, measurements of the effective characteristics of these patterns have shown them to be less variable than a nondisplay movement. The relationship between mean position of limb and the standard deviations of the movements measured in this study is shown in Fig. 5. The difference in relative variability for display and nondisplay movements is clear. The regression line for the display movements is negative ($b_1 = -0.1$), while the line for the nondisplay movements is positive ($b_2 = +0.2$). This negative slope for the display movements indicates very effective selection against variability, i.e., selection for stereotypy.

Variability measures of ethologically interesting behavior patterns are rare. However, Salmon (1967) and Salmon and Atsaides (1968) provide data on six species of *Uca*. Their measurements were only on the time (number of frames of film) taken to complete waves of the major cheliped of a male *Uca*, or other temporal measures. It is not known if such temporal characteristics are important to the crabs. Comparison of movement duration variability in the two groups of crustaceans shows quite similar levels of variation. When females were absent, wave duration in *Uca rapax* averaged 3.6 sec, with a standard deviation of 1.9 sec (Salmon, 1967). This is quite similar to the numbers for the cheliped presentation display of *Clibanarius vittatus*, which averaged 3.5 frames with a standard deviation of 1.4. The waves of *U. mordax* (Salmon, 1967) averaged 6.6 sec, with a standard deviation of 2.6. which is just a little more variable than the ambulatory raise display in *C. vittatus*, which averaged 5.3 with a SD of 1.9. *U. virens* (average 2.8. SD 0.80) and *U. longisignalis* (average 3.7, SD 1.0) (Salmon and Atsides, 1968) would appear to be less variable in movement duration than *C. vittatus*, although *F* ratios indicated that the differences are not significant.

The similarity of variability in movement duration is more striking when one considers that the movements may be serving different functions in the two groups of Crustacea (while definitely agonistic–territorial, the waving of *Uca* species may also serve as a sexual signal). When one compares these values with some of those available for vertebrates, the similarities are even more surprising, although some differences appear. Dane *et al.* (1959) analyzed the displays of the goldeneye duck, *Bucephala clangula*. They obtained temporal variability measures for a number of movements [again without evidence that duration of movement is important to the ducks, although sequence analysis was carried out showing the communicatory effect of certain movements (Dane and van der Kloot, 1964)]. For flock displays of male goldeneyes, the variances are very similar to those seen in hermit and fiddler crab displays: "nodding" averaged 2.7 sec, SD 1.9; "masthead" averaged 4.80, SD 2.5. However, some flock displays (such as "head throw–kick," average 1.2 SD 0.10) and most of the precopulatory displays of the male ("head flick," 0.23, SD 0.05; "head rubbing," 0.75, SD 0.08, "display drinking," 2.02, SD 0.11) are clearly less variable than the crustacean movements, with the exception of the waves of *Uca speciosa* (Salmon, 1967), which averaged 0.5 sec with a standard deviation of 0.01.

The duration variability for the goldeneye duck and six species of fiddler crabs, the display movements of *Clibanarius vittatus*, and song duration measurement for four species of birds are presented in Table VII. The bird song data are included to compare territorial behaviors in some crustaceans and vertebrates, although the mixing of sensory modalities (visual–auditory) makes any conclusions tenuous. The most important statistic for each movement is the coefficient of variation ($\bar{V}$), which is the variability measure most reliably compared across various samples. In Table VIII, F ratios and associated probability values for comparisons of the average $\bar{V}$ values for the different animals are presented. The F ratios must be viewed with caution, since some

Table VII. Measurements of Movement Duration Variability for Different Communicatory Behavior Patterns[a]

Species and pattern	N	Mean	SD	$\bar{V}$	$\bar{X}_{\bar{V}}$	Reference
Goldeneye duck						
Masthead	13	4.80	2.5	52.09		
Nodding	14	2.7	1.9	70.39		Dane *et al.* (1959)
Head throw–kick	38	1.2	0.1	8.33		
					28.09	
Head flick	11	0.23	0.05	21.74		
Head rubbing	18	0.75	0.08	10.67		
Display drinking	31	2.02	0.11	5.45		
Fiddler crab waving						
Uca speciosa	60	0.5	0.01	2.00		
U. rapax	60	3.6	1.9	52.77		Salmon (1967),
						Salmon and Atsaides
U. virens	90	2.8	0.8	28.57		
					29.51	(1968)
U. longisignalis	38	3.7	1.0	27.03		
U. mordax	60	6.6	2.6	39.69		
U. pugnax	60	3.7	1.0	27.03		
Hermit crab, *C. vittatus*						
Ambulatory raise	19	0.31	0.11	35.48		
Cheliped presentation	34	0.21	0.10	47.62	41.28	
Cheliped extension	14	0.27	0.11	40.74		
Bird song						
Mexican junco	67	1.63	0.29	17.8		
Oregon junco	77	1.49	0.20	13.6		Heckenlively (1970)
					20.3	
Brown towhee	61	1.52	0.31	20.4		
Black-throated Sparrow	67	1.02	0.31	30.4		

[a] All means and standard deviations are in seconds.

Table VIII. *F* Ratio Tests on $\bar{V}$ for Duration of Movements[a]

Coefficients compared	*F* ratio	*p*
All duck displays—all hermit crab	1.47	>0.10
All *Uca* displays—all hermit crab	1.42	>0.10
All bird song—all hermit crab	2.03	<0.05
Black-throated sparrow—all hermit crab	1.40	>0.10
Precopulatory duck display—hermit crab	3.59	<0.01
Uca speciosa—other *Uca* species	17.00	<0.01
Uca speciosa—precopulatory duck display	5.70	<0.01
Uca speciosa—Display drinking of duck	2.72	<0.01
Territorial—sexual (duck)	1.13	>0.10
Territorial—agonistic (hermit crab)	1.62	>0.10
All bird behaviors—all crustacean behaviors	1.47	>0.10

[a] Data are from Table VII.

of the necessary condition for running such tests (Kerfoot, 1969; Lewontin. 1966) probably are not satisfied (and I cannot check these from the published data). In addition to statistical questions, it should be remembered that time variability may not be important in all or any of the animals studied.

Keeping in mind the "tenuous legality" of the probability values, we can see from Table VIII that considering all duck movements listed or all *Uca* species measured, the variability is the same and neither is different from the bird song data or from hermit crab display duration. The average bird song is, however, significantly less variable than the average *C. vittatus* display. There is no difference for the black-throated sparrow. If an average of only those goldeneye duck movements which are directly precopulatory (not flock displays) is compared to *C. vittatus* agonistic displays. the former are less variable. Clearly, *Uca speciosa* is very unusual in the degree of duration stereotypy. [This was not true for the sounds produced by *Uca speciosa* compared to those of *Uca rapax* (Salmon, 1967).] While making these somewhat tenuous comparisons, I also lumped fiddler crab and bird song values to get a $\bar{V}$ for territorial behaviors. This was not significantly different from purely sexual behaviors (goldeneye duck) or more purely agonistic behaviors (Hermit crabs). And as a final comparison, all crustacean values were averaged ($\bar{V} = 35.4$) and compared to the average for all bird data ($\bar{V} = 24.1$). No significant difference was found. It should be pointed out that for hermit crab displays, position variability ($\bar{V} = 7.5$) is very significantly ($F = 5.5$, $p < 0.01$) smaller than duration variability ($\bar{V} = 41$).

It would now be of interest to be able to compare the differences in variability of display and nondisplay movements which clearly serve analogous functions in different groups of animals. Species identification could require less ambiguous signals that those showing agonistic level (also see Marler, 1957). Also, species identification patterns could be more stereotyped in an area of congeneric sympatry compared to a geographic area where there is only one member of a taxon. And perhaps there are differences in signal development from groups to group that could tell us something about modes of neuronal organization. Variability measures, like sequence analysis, require more work than the (needed) initial ethogram. But particularly with the increased availability of computers for analysis of data, such measures should be made.

ACKNOWLEDGMENTS

This work was supported in part by a research grant from the Rackham School of Graduate Studies, University of Michigan, by a postdoctoral fellowship (MH-14, 274) from the National Institutes of Health, and by a grant (HD 02847) from the National Institutes of Health. Thanks are given to Steve Arnold, Hugh Dingle, Bruce Oakley, and Michael Salmon for their comments on the manuscript.

REFERENCES

Altevogt, R., 1955, Beobachtungen und Untersuchungen an indischen Winkerkrabben, *Z. Morphol. Ökol. Tiere* **43**: 501–522.

Altevogt, R., 1957, Untersuchungen zur Biologie, Ökologie und Physiologie indischer Winkerkrabben, *Z. Morphol. Ökol. Tiere* **46**: 1–110.

Altmann, S., 1965, Sociobiology of rhesus monkeys. II. Stochastics of social communication, *J. Theoret. Biol.* **8**: 490–522.

Ball, E. E., Jr., 1968, Activity patterns and retinal pigment migration in *Pagurus* (Decapoda, Paguridea), *Crustaceana* **14**: 302–306.

Beer, C. G., 1959, Notes on the behaviour of two estuarine crab species, *Tr. Royal Soc. New Zealand* **86**: 197–203

Bethe, A. D., 1897, Das Nervensystem von *Carcinus maenas*, *Arch. Mikro. Anat.* **50**: 462–544; 590–639.

Blest, A. D., 1963, The concept of "ritualization," *in* "Current Problems in Animal Behavior" (W. H. Thorpe and O. L. Zangwill, eds.) pp. 102–124, Cambridge University Press, London.

Bovbjerg, R. V., 1960, Courtship behavior of the lined shore crab, *Pachygrapsus crassipes* Randall, *Pac. Sci.* **14**: 421–422.

Braestrup, F. W., 1966, Social and communal display, *Phil. Tr. Royal Soc. Lond.* **251**: 375–386.

Brock, F., 1927, Das Verhalten des Einsidlerkrebses *Pagurus arrosor* Herbst während des Aufsuchens, Ablösens und Aufpflanzens der Seerose *Sagartis parasitica* Goose, *Wilhelm Roux Arch. Entwick. Organ.* **112**: 204–238.

Cameron, A. M., 1966, Some aspects of the behaviour of the soldier crab, *Mictyrus longicarpus*, *Pac. Sci.* **20**: 224–234.

Cott, H. B., 1929, Observations on the natural history of the land-crab *Sesarma meinerti*, from Beira, with special reference to the theory of warning colours, *Proc. Zool. Soc. Lond.* **1929**: 679–692.

Crane, J., 1941, Crabs of the genus *Uca* from the west coast of Central America, *Zoologica* **26**: 145–208.

Crane, J., 1943, Display, breeding and relationships of fiddler crabs (Brachyura, genus *Uca*) in the Northeastern United States, *Zoologica* **28**: 217–223.

Crane, J., 1957, Basic patterns of display in fiddler crabs (Ocypodidae, genus *Uca*), *Zoologica* **42**: 69–82.

Crane, J., 1966, Combat, display and ritualization in fiddler crabs (Ocypodidae, genus *Uca*), *Phil. Tr. Royal Soc. Lond.* **251**: 459–472.

Crane, J., 1967. Combat and its ritualization in fiddler crabs (Ocypodiadae) with special reference to *Uca rapax* (Smith), *Zoologica* **52**: 49–76.

Daanje, A., 1950, On locomotory movements in birds and the intention movements derived from them, *Behaviour* **3**: 48–98.

Dane, B., and van der Kloot, W., 1964, An analysis of the display of the goldeneye duck (*Bucephala clangula* (L.)), *Behaviour* **22**: 282–328.

Dane, B., Walcott, C., and Drury, W. H., 1959, The form and duration of the display actions of the goldeneye (*Bucephala clangula*), *Behaviour* **14**: 265–281.

Darwin, C., 1872, "The Expression of the Emotions in Man and Animals," University of Chicago Press, Chicago.

Daumer, K., Jander, R., and Waterman, T. H., 1963, Orientation of the ghost-crab *Ocypode* in polarized light, *Z. vergl. Physiol.* **47**: 56–76.

Dijkgraaf, S., 1955, Lauterzeugung und Schallwahrnehmung bei der Languste (*Palinurus vulgaris*), *Experientia* **11**: 330–331.

Dingle, H., 1969, A statistical and information analysis of aggressive communication in the mantis shrimp *Gonodactylus bredini* Manning, *Anim. Behav.* **17**: 561–575.

Dingle, H., and Caldwell, R. L., 1969, The aggressive and territorial behaviour of the mantis shrimp *Gonodactylus bredini* Manning (Crustacea: Stomatopoda), *Behaviour* **33**: 115–136.

Dybern, B. I., and Höisaeter, T., 1965, The burrows of *Nephrops norvegicus* (L.), *Sarsia* **21**: 49–55.

Fish, J. F., 1966, Sound production in the American lobster, *Homarus americanus* H. Milne Edwards (Decapoda Reptantia), *Crustaceana* **11**: 105–106.

Griffin, D. J. G., 1965, The behaviour of shore crabs, *Austral. Nat. Hist.* **15**: 87–91.

Griffin, D. J. G., 1968, Social and maintenance behaviour in two Australian ocypodid crabs (Crustacea: Brachyura), *J. Zool. Lond.* **156**: 291–305.

Guinot-Dumortier, D., and Dumortier, B., 1960, La stridulation chez les crabes, *Crustaceana* **1**: 117–155.

Hazlett, B. A., 1966*a*, Social behavior of the Paguridae and Diogenidae of Curaçao, *Studies Fauna Cuaçao* **23**: 1–143.

Hazlett, B. A., 1966*b*, The behavior of some deep-water hermit crabs (Decapoda: Paguridea) from the straits of Florida, *Bull. Mar. Sci.* **16**: 76–92.

Hazlett, B. A., 1966*c*, Observations on the social behavior of the land hermit crab, *Coenobita clypeatus* (Herbst), *Ecology* **47**: 316–317.

Hazlett, B. A., 1967, Interspecific shell fighting between *Pagurus bernhardus* and *Pagurus cuanensis* (Decapoda, Paguridea), *Sarsia* **29**: 215–220.

Hazlett, B. A., 1968*a*, The sexual behavior of some European hermit crabs, *Pubbl. Staz. Zool. Napoli* **36**: 238–252.

Hazlett, B. A., 1968*b*, The phyletically irregular social behavior of *Diogenes pugilator* (Anomura, Paguridea), *Crustaceana* **15**: 31–34.

Hazlett, B. A., 1968*c*, Size relationships and aggressive behavior in the hermit crab *Clibanarius vittatus*, *Z. Tierpsychol.* **25**: 608–614.

Hazlett, B. A., 1968*d*, Communicatory effect of body position in *Pagurus bernhardus* (L.) (Decapoda, Anomura), *Crustaceana* **14**: 210–214.

Hazlett, B. A., 1968*e*, Stimuli involved in the feeding behavior of the hermit crab *Clibanarius vittatus* (Decapoda, Paguridea), *Crustaceana* **15**: 305–311.

Hazlett, B. A., 1969, Further investigations of the cheliped presentation display in *Pagurus bernhardus* (Decapoda, Anomura), *Crustaceana* **17**: 31–34.

Hazlett, B. A., 1970*a*, Tactile stimuli in the social behavior of *Pagurus bernhardus* (Decapoda, Pagridae), *Behaviour* **36**: 20–48.

Hazlett, B. A., 1970*b*, The effect of shell size and weight on the agonistic behavior of a hermit crab, *Zeit. f. Tierpsychol.* **27**: 369–374.

Hazlett, B. A., and Bossert, W. H., 1965, A statistical analysis of the aggressive communications systems of some hermit crabs, *Anim. Behav.* **13**: 357–373.

Hazlett, B. A., and Bossert, W. H., 1966, Additional observations on the communications systems of hermit crabs, *Anim. Behav.* **14**: 546–549.

Hazlett, B. A., and Winn, H. E., 1962*a*, Sound production and associated behavior of Bermuda crustaceans (*Panulirus*, *Gonodactylus*, *Alpheus* and *Synalpheus*), *Crustaceana* **4**: 25–38.

Hazlett, B. A., and Winn, H. E., 1962*b*, Characteristics of a sound produced by the lobster *Justitia longinanus*, *Ecology* **43**: 741–742.

Heckenlively, D. B., 1970, Song in a population of black-throated sparrows, *Condor* **72**:24–36.

Hiatt, R. W., 1948, The biology of the lined shore crab, *Pacygrapsus crassipes* Randall, *Pac. Sci.* **2**: 135–213.

Huxley. J., 1923, Courtship activities in the red-throated diver (*Colymbus stellatus* Pontopp.); together with a discussion on the evolution of courtship in birds, *J. Linn. Soc.* **35**: 253–293.

Huxley, J., 1966, Introduction to a discussion of ritualization of behaviour in animals and man, *Phil. Tr. Royal Soc. Lond.* **251**: 249–271.

Johnson, V. R., Jr., 1969, Behavior associated with pair formation in the banded shrimp *Stenopus hispidus* (Olivier), *Pac. Sci.* **23**: 40–50.

Kerfoot, W. C., 1969, Selection of an appropriate index for the study of the variability of lizard and snake body scale counts, *Systematic Zool.* **18**: 53–62.

Knudsen, J. W., 1960, Aspects of the ecology of the California pebble crabs (Crustacea: Xanthidae), *Ecol. Monographs* **30**: 165–185.

Lewontin, R. C., 1966, On the measurement of relative variability, *Systematic Zool.* **15**: 141–142.

Limbaugh, C., Pederson, H., and Chase, F. A., Jr., 1961, Shrimps that clean fishes, *Bull. Mar. Sci.* **11**: 237–257.

Lindberg, R., 1955, Growth, population dynamics, and field behavior in the spiny lobster, *Panulirus interruptus* (Randall), *Univ. Calif. Publ. Zool.* **59**: 157–248.

Linsenmair, K. E., 1965, Optische Signalisierung der Kopulationshöhle bei der Reiterkrabbe *Ocypode saratan* Forsk. (Decapoda Brachyura Ocypodidae), *Naturwissenschaften* **52**: 256–257.

Linsenmair, K. E., 1967, Konstruktion und Signalfunktion der Sandpyramide der Reiterkrabbe *Ocypode saratan* Forsk. (Decapoda Brachyura Ocypodidae), *Z. Tierpsychol.* **24**: 403–456.

Lorenz, K., 1950, The comparative method in studying innate behaviour patterns, *Symp. Soc. Exp. Biol.* **4**: 221–269.

Magnus, D., 1960, Zur Ökologie des Landeinsiedlers *Coenobita jousseaumei* Bouvier und der Krabbe *Ocypode aegyptiaca* Gerstaecker am Roten Meer, *Verhandl. Deutsch. Zool. Ges.* **54**: 316–329.

Marler, P., 1957, Specific distinctivenes in the communication signals of birds, *Behaviour* **11**: 13–39.

Morris, D., 1956, The feather postures of birds and the problem of the origin of social signals, *Behaviour* **9**: 75–114.

Morris, D., 1957, "Typical intensity" and its relationship to the problem of ritualization, *Behaviour* **11**: 1–12.

Morris, D., 1966, The rigidification of behaviour, *Phil. Tr. Royal Soc. Lond.* **251**: 327–330.

Moulten, J. M., 1957, Sound production in the spiny lobster *Panulirus argus* (Latreille), *Biol. Bull.* **113**: 286–295.

Moynihan, M., 1960, Some adaptations which help to promote gregariousness, *in* "Proc. XII Internat. Ornith. Congr.," Helsinki, pp. 523–541.

Provenzano, A. J., Jr., 1959, The shallow-water hemit crabs of Florida, *Bull. Mar. Sci.* **9**: 349–420.

Reese, E. S., 1962, Submissive posture as an adaptation to aggressive behavior in hermit crabs, *Z. Tierpsychol.* **19**: 645–651.

Reese, E. S., 1964, Ethology and marine zoology, *Oceanograph. Mar. Biol. Ann. Rev.* **1964**: 455–488.

Ross, D. M., and Sutton, L., 1961, The association between the hermit crab *Dardanus arrosor* (Herbst) and the sea anemone *Calliactis parasitica* (Couch), *Proc. Royal Soc. B* **155**: 282–291.

Ross, D. M., and Sutton, L., 1968, Detachment of sea anemones by commensal hermit crabs and by mechanical and electrical stimuli, *Nature* **217**: 380–381.

Ryan, E. P., 1966, Pheromone: Evidence in a decapod crustacean, *Science* **151**: 340–341.

Salmon, M., 1965, Waving display and sound production in the courtship behavior of *Uca pugilator*, with comparisons to *U. minax* and *U. pugnax*, *Zoologica* **50**: 123–150.

Salmon, M., 1967, Coastal distribution, display and sound production by Florida fiddler carbs (genus *Uca*), *Anim. Behav.* **15**: 449–459.

Salmon, M., and Atsaides, S. P., 1968, Behavioral, morphological and ecological evidence for two new species of fiddler crabs (Genus *Uca*) from the Gulf Coast of the United States, *Proc. Biol. Soc. Wash.* **81**: 275–290.

Salmon, M., and Atsaides, S. P., 1969, Sensitivity to substrate vibration in the fiddler crab, *Uca pugilator* Bosc, *Anim. Behav.* **17**: 68–76.

Salmon, M., and Stout, J. F., 1962, Sexual discrimination and sound production in *Uca pugilator* Bosc, *Zoologica* **47**: 15–20.

Schöne, H., 1961, Complex behavior, *in* "The Physiology of Crustacea" (T. H. Waterman, ed.) Vol. II, pp. 465–520, Academic Press, New York.

Schöne, H., 1968, Agonistic and sexual display in aquatic and semiterrestrial brachyuran crabs, *Am Zoologist* **8**: 641–654.

Schöne, H., and Eibl-Eibesfeldt, I., 1965, *Grapsus grapsus* (Brachyura) Drohen, *Encylop. Cinemat. E* **599**: 1–8.

Schöne, H., and Schöne, H., 1963, Balz und andere Verhaltenweisen der Mangrovekrabbe *Goniopsis cruentata* Latr. und das Winkverhalten der eulitoralen Brachyuren, *Z. Tierpsychol.* **20**: 641–656.

Selous, E., 1901, *Bird watching*, Dent, London.

Smith, W. J., 1965, Message, meaning. and context in ethology, *Am. Naturalist* **99**: 405–409.

Tinbegen, N., 1952, "Derived" activities; their causation, biological significance, origin, and emancipation during evolution, *Quart. Rev. Biol.* **27**: 1–32.

Tinbergen, N., 1964, The evolution of signaling devices, *in* "Social Behavior and Organization among Vertebrates" (W. Etkin, ed.) pp. 206–230, Chicago Press, Chicago.

von Hagen, H.-O., 1961, Experimentelle Studien zum Winken von *Uca tangeri* in Südspanien, *Verhand. Deutsch. Zool. Ges.* **55**: 424–432.

von Hagen, H.-O., 1968, Zischende Drohgeräusche bei westindischen Krabben, *Naturwissenschaften* **55**: 139.

Wright, H. O., 1968, Visual displays in brachyuran crabs: Field and laboratory studies, *Am. Zoologist* **8**: 655–665.

Zajonc, R. B., 1965, Social facilitation, *Science* **149**: 269–274.

Chapter 4

AGGRESSIVE BEHAVIOR IN STOMATOPODS AND THE USE OF INFORMATION THEORY IN THE ANALYSIS OF ANIMAL COMMUNICATION

Hugh Dingle

Department of Zoology
University of Iowa
Iowa City, Iowa

I. INTRODUCTION

The study of animal communication is one of the most active areas within the discipline of ethology. Yet in spite of a seemingly ever-accelerating rate of publication, there is still a paucity of quantitative studies. The typical analysis of communication presents a description of certain stereotyped movements or "displays" and attributes a communicatory function to these on the basis of observed or presumed changes in behavior. However, until quantitative studies are made of such movements, and changes in frequency as a result of interactions are noted, no function can be determined with certainty. Part of the problem with quantitative analysis has been that no single method has been applicable to all kinds of behavior, and the search for appropriate analysis for a given situation is often tedious. There are, nevertheless, some interesting studies available (e.g., Baerends *et al.*, 1955; Dane *et al.*, 1959; Wiepkema, 1961; Mittelstaedt, 1964; Stokes, 1962; Nelson, 1964, 1965; Shaw, 1968; see also Hinde, 1966), and the reader should refer to these for descriptions and discussions of the various methodologies.

One quantitative approach particularly applicable to the study of communication is information theory. But, as indicated, quantitative studies are not common in the literature of behavior; this is especially true with respect to analyses utilizing information theory. Five papers which make use of this approach are those of Haldane and Spurway (1954) on information apparently transferred in the dance communication system of the honey bee, Wilson (1962) on chemical communication in the fire ant, Hazlett and Bossert (1965) on visual communication in the aggressive interactions of hermit crabs, Dingle (1969) on visual and tactile communication in the aggressive interactions of the stomatopod crustacean *Gonodactylus bredini*, and Altmann (1965) on the various types of communication involved in the social behavior of rhesus monkeys.

As will be indicated in this chapter, some of the comparisons between the various species are potentially of considerable interest. These studies indicate that information theory deserves more attention from ethologists; it has the particular advantage of allowing calculation of values which, since they have the same units, can be compared across species no matter what the type of behavior studied (aggressive, sexual, territorial, etc.) or the means of communication used (visual, chemical, sound, etc.). From these comparisons it may be possible to make general statements concerning communication, as, for example, the relation between the information content of a series of displays and the type of display pattern or between rate of transmission and the sensory modality involved. The data necessary to calculate information units can also be used to study other facets of behavior such as distribution or sequential nature of displays (Hazlett and Bossert, 1965; Dingle, 1969).

This chapter will present a brief introduction to information theory with special reference to its use in the analysis of communication, followed by an example of its use in the analysis of the aggressive communication of the stomatopods *Gonodactylus bredini* and *G. spinulosus*. The presentation is aimed at the mathematically unsophisticated reader who may, nevertheless, wish to consider the possibilities of applying information theory in his own work; I apologize in advance to those who may find it, as a result, too elementary. More comprehensive and sophisticated treatments of information theory may be found in Shannon (1948), Shannon and Weaver (1949), Quastler (1958*a*), Brillouin (1960), and Garner (1962).

II. INFORMATION THEORY

A. The Measure of Information

The unit of measurement most frequently used for information is the *bit* (a contraction of binary digit). Briefly, a bit is the amount of uncertainty eliminated by answering a single yes–no question, where the two answers are equally probable, often symbolized by the binary digits 0 and 1. To take the simplest possible case, suppose the males and females of a species

Fig. 1. Pattern of eight stones with shrimp captured under stippled stone (see text).

of shrimp can be distinguished morphologically and the sex ratio is 1 : 1. On capturing an animal, we examine it and find it is a female. The information present in this event is precisely one bit, because we have answered the yes–no question "Is this animal a female?" and the two answers were equally probable. The number of bits present in a given situation equals the power to which 2 must be raised to equal the total number of equally probable alternatives. One can also use powers of 10 or of e; if the base e is used, information is expressed in *nits*, while the unit in the base 10 system is the Hartley in honor of the mathematician R. V. L. Hartley, who was first to consider in detail the problem of measuring information. The measure of information is usually given the symbol H, so that if m is the number of equally likely alternatives, $m = 2^H$, which can also be written $H = \log_2 m$. Thus in sexing our shrimp $m = 2$ and H, therefore, equals one bit.

Now let us further suppose that a shrimp was captured under one of eight stones as shown in Fig. 1 with the stone in question shaded, and we wish to identify the particular stone. We can do this by asking three binary yes–no questions as follows:

1. Is it one of the four stones on the left? (No or 0)
2. Is it one of the two stones on the far right? (No or 0)
3. Of the two stones remaining, is it the upper one? (Yes or 1)

Since we can locate the stone with three questions, there are three bits of information present. In this case we have $H = \log_2 8 = 3$ bits, since there are eight equally probable alternatives (stones). If all m alternatives are equally probable, the probability (p) of one alternative is $1/m$ and m is therefore equal to $1/p$. Thus the formula for H can also be written $H = \log_2(1/p)$.

Suppose, however, we face a situation in which alternatives are not equally probable. Consider a hypothetical case in which an ethologist presents a painted model to each of several males of a species and the population of males either attacks or flees from the model in the proportion 80% attack to 20% flee. Therefore, $p\,(\text{attack}) = 0.80$ and $p\,(\text{flee}) = 0.20$. In view of the ratio between attacking and fleeing, we can predict that for any presentation a given male is likely to attack, and, hence, when an attack occurs, we do not obtain much information. Fleeing is rarer, so its occurrence yields a greater information gain. For the information present in the choice of a particular alternative i, we calculate $h_i = \log_2 (1/p_i)$. Thus

$$h\text{ (attack)} = \log_2 (1/0.80) = \log_2 1.25 = 0.32 \text{ bit}$$
$$h\text{ (flee)} = \log_2 (1/0.20) = \log_2 5.00 = 2.32 \text{ bits}$$

The formula yields different values of information depending on whether a particular male attacks or flees the model. With many replications of the experiment, we can calculate a value (H) which would be the mean amount of information present per model presentation. To do this, we must weight the equation appropriately; for example, for each attack we obtain 0.32 bit, but attacks occur 80% of the time. Thus the mean information available is

$$H = 0.80\ (0.32) + 0.20\ (2.32) = 0.72 \text{ bit}$$

Note that H in this case is less than the 1 bit obtained when we were sexing shrimp. There, with two equally probable events (male and female), we had $H = 0.5\ (1) + 0.5\ (1) = 1$ bit.

When there are m alternatives appropriately weighted, the calculation can be stated formally as follows:

$$H = p_1h_1 + p_2h_2 + \ldots + p_mh_m \tag{1}$$

or more succinctly

$$H = \sum p_ih_i \tag{2}$$

which, since $h_i = \log_2 (1/p_i)$, can also be written

$$H = \sum p_i \log_2 (1/p_i) \tag{3}$$

usually expressed as

$$H = -\sum p_i \log_2 p_i \tag{4}$$

since $\log_2 (1/p_i) = -\log_2 p_i$. Equation (4) is often referred to as the Shannon–Wiener formulation for information present, being named for the founders of modern communication theory (Shannon) and cybernetics (Wiener). This formula is used for estimating information present (uncertainty) when the population being sampled is effectively infinite (Pielou, 1966) and is the one most applicable to animal communication. Usually, p_i is taken to be n_i/N where N is the total number of events sampled and n_i is the number falling into the ith of m categories ($\sum n_i = N$). The papers of Miller (1955) and Pielou (1966) should be consulted for discussion of the bias present in using n_i/N to estimate p_i.

As we saw previously, when there are m equally probable alternatives, $H = \log_2 m$; this is the maximum possible value of H and as such is frequently designated H_{max}. This value depends only on the number of alternatives or categories and is therefore a property of how categories are determined; for example, a behavior analysis that included an animal's every movement would contain more categories and hence a greater H_{max} than an analysis which lumped behavior into displays. This causes real problems in the analysis of behavior, and these will be discussed in more detail later. The ratio H/H_{max} in a given situation is called the relative uncertainty or relative information present; it is a measure of the extent to which the possible diversity present in a given set of alternatives is utilized to carry information. That portion of the possibilities not involved in carrying information is called "redundancy" (now usually symbolized R) and is the complement of the relative uncertainty. Thus $R = 1 - H/H_{max}$ (Quastler, 1958*a*), and redundancy is zero when all possibilities occur with equal frequency ($H = H_{max}$). The classic example of redundancy in the English language is the sequence of letters "qu", where given the letter "q" we can predict with virtually complete certainty that it will be followed by "u". An example from behavior is the invariable sequence of egg-laying by the female followed immediately by fertilization by the male occurring in many animals. In behavior, redundancy may be important even though it represents a reduction in the amount of information carried. It can be a powerful means of reducing errors of interpretation (noise) and thus of increasing the effectiveness of communication. Wherever accuracy is a prime necessity in biological communication, one can predict that natural selection will have ensured the incorporation of redundancy into the signals. Many cases of courtship behavior where certain displays or sequences of displays are highly repetitive are undoubtedly examples of the extensive incorporation of redundancy, through natural selection, into a system of communication requiring accurate interpretation. Repetition of acts which serve, in part, to physiologically prime a recipient is a separate problem.

The theoretical relationship between information, redundancy, and noise is summarized by Dancoff's principle (Dancoff and Quastler, 1953; Quastler, 1958*b*), which deals with the "economics" of information. In any system, both errors (noise) and redundancy are costly. The optimum amount of redundant information does not necessarily eliminate errors but rather minimizes the sum: cost of errors plus cost of redundancy plus cost of error checking. Dancoff's principle predicts with regard to living things that an organism which has undergone natural selection will approach such an optimum. In other words, it will make as many errors as it can get away with and do so with the minimum redundancy needed; this effectively limits the amount of redundant information present.

B. Measures of Communication

In information theory, communication is used in a broad sense to mean any relation between variables, using whatever means, whether or not intentional, resulting in the mutual reduction of uncertainty (Quastler, 1958*a*).

Table I. Data from Hypothetical Behavioral Interactions Involving Two Animals

	Following act (B)				
Initial act (A)	$j =$ (1) attack	(2) flee	(3) threat	Totals (n_i)	p_i
$i =$ (1) Attack	10	25	5	40	0.40
(2) Flee	25	5	10	40	0.40
(3) Threat	10	5	5	20	0.20
Totals (n_j)	45	35	20	$100 = n$	
p_j	0.45	0.35	0.20		1.00

Note that this is a somewhat different meaning of the term from that in everyday usage, where some sort of purposiveness or intent is usually implied. In a behavioral analysis we are most concerned with the relationships between two variables, as when one animal performs a display or act and a second animal responds to it.

A hypothetical case of two related behavioral variables is presented in Table I. In this situation two animals are placed in a container, and the occurrence of three acts, "attack," "flee," and "threat," is recorded. The animal performing first is arbitrarily designated A, and its act the initial act, act i. Similarly, the responding animal is designated B, and its response is the following act, j. The body of the table contains the numbers of times the specified following acts occurred after given initial acts; the probability of occurrence of a given two-act sequence is designated p_{ij} (e.g., the probability of the sequence attack–attack is 10/100 or 0.10). Also included in the table are the p_i's and p_j's which are the overall probabilities of occurrence of the acts i and j, respectively.

From these data it is possible to calculate various values of information present; thus

$$H_A = -\sum p_i \log_2 p_i = 1.522 \text{ bits} \tag{5}$$

for the information present in the overall distribution of the initial acts, and

$$H_B = -\sum p_j \log_2 p_j = 1.513 \text{ bits} \tag{6}$$

for following acts, and

$$H_{A,B} = -\sum p_{ij} \log_2 p_{ij} = 2.861 \text{ bits} \tag{7}$$

for the joint information present.

It is also possible from the data in Table I to calculate values for "conditional information present." This is the information present in the acts performed by, say, animal B if the acts performed by A are known. The conditional information present is designated $H_{B/A}$ [sometimes written

$H_A(B)$] and is calculated from the conditional probabilities, $p_{j/i}$ [also written $p_i(j)$], which are the probabilities of animal B performing act j given that animal A has just performed act i (e.g., in Table I given that an attack has just occurred, the probability that the next act will be an attack is 10/40 or 0.25). The equation for this calculation can be written in several ways (see Quastler, 1958*a*), but the simplest is probably the following:

$$H_{B/A} = -\sum p_{ij} \log_2 p_{j/i} \tag{8}$$

where p_{ij}'s are again the joint probabilities. Similarly,

$$H_{A/B} = -\sum p_{ij} \log_2 p_{i/j} \tag{9}$$

which is the conditional information present in the initial acts performed by animal A, once the following acts are known.

Next one can calculate information transmitted or communicated [symbolozed H_t or $T\,(A;B)$]. The relation between H_t and the other information functions is

$$H_t = H_A - H_{A/B} = H_B - H_{B/A} \tag{10}$$

which indicates the gain in certainty about animal A's acts from observing animal B's acts, and vice versa. H_t is thus the mutual reduction in uncertainty, and a measure of the internal constraints in a communication system (Quastler, 1958*a*). It can also be calculated from the joint uncertainty. Thus

$$H_t = H_A + H_B - H_{A,B} \tag{11}$$

This is the most convenient method if it is not necessary to compute conditional information present. With respect to our hypothetical behavioral interaction given in Table I, $H_t = 1.522$ bits $+ 1.513$ bits $- 2.861$ bits $=$ 0.174 bit, the amount of information transmitted per act. This represents the restriction of B's behavioral acts by A's acts and can be considered the amount of information that B indicates to the observer by his actions that he has received from A. It is thus a minimum estimate of the information transmitted per act (Hazlett and Bossert, 1965; Dingle, 1969). The relationships between the various information functions discussed above are indicated graphically in Fig. 2.

In the case of Table I, we can also compute information transmitted per interaction per individual, H_t/IN (Hazlett and Bossert, 1965; Dingle, 1969), which allows us to look at the average information in single behavioral exchanges between individuals. To do this requires a knowledge of the number of acts taking place during the average interaction between two animals. For the situation in question we shall assume an average of three acts by each animal per interaction. Multiplying this figure times the value of bits for H_t gives 0.522 bit as the value for H_t/IN. From this value, plus estimates of the mean duration of an interaction, one can calculate the rate of trans-

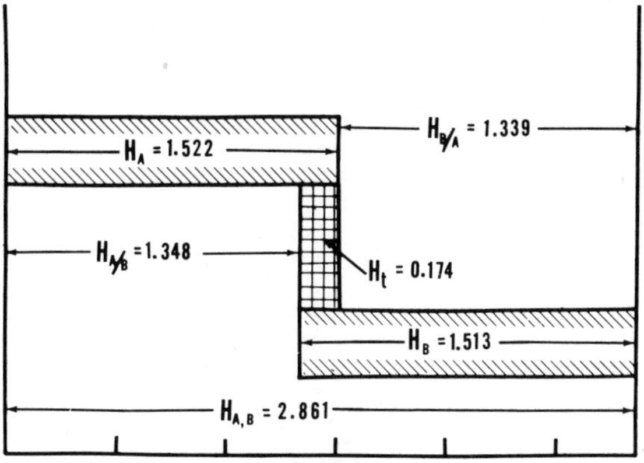

Fig. 2. The relationships between the various values of H. The figures refer to the values (in bits) calculated from the "data" given in Table I.

mission per unit time. If we assume an average interaction duration of 10 sec for our hypothetical situation, then the rate of transmission is 0.052 bit/sec. Since temporal factors can be important in communication, it is interesting to compare these rates across different types and modes of transmission.

Returning to our original discussion of information as a function of a series of binary yes–no questions, we can state the meaning of each of the functions discussed in terms of such questions. The terms (after Quastler, 1958*a*) usually used to designate these functions are in parentheses. Thus

H_A = number of questions (or operations) specifying A (uncertainty or information of source)

H_B = number of questions specifying B (uncertainty of destination)

$H_{A,B}$ = number of questions specifying the whole system (uncertainty of the system)

$H_{A/B}$ = number of questions specifying A if B is given (equivocation)

$H_{B/A}$ = number of questions specifying B if A is given (ambiguity)

H_t = number of questions which apply to the specification of both A and B (information transmitted or communicated)

C. Some Comments on Terminology

As in any developing field, especially one which spans several disciplines, information theory has had problems with terminology. On occasion

Table II. Various Symbols and Terms Used in Information Theory[a]

Symbol	Definition	Term	Other symbols	Other terms
H	$-\sum p_i \log_2 p_i$	Information present	U	Entropy, uncertainty
H_t	$H_A + H_B - H_{A,B}$	Information transmitted	T, R	Reduction in transition uncertainty
$\frac{H}{H_{\mathbf{max}}}$	$\frac{-\sum p_i \log_2 p_i}{\log_2 m}$	Relative information Present	R	Relative entropy or uncertainty
R	$1 - \frac{H}{H_{\mathbf{max}}}$	Redundancy	S	Stereotypy

[a] First three columns give usages in this paper.

those in different fields have assigned different terms or symbols to the various values of information, and terminology has changed as parameters have been more precisely defined and better understood or when there has been a need for modification in order to ensure consistency and avoid confusion. For example, Shannon (1948) originally assigned no symbol to redundancy but gave information transmitted the symbol R. More recently, R has been used for redundancy and T or H_t for transmission, although Altmann (1965) calls redundancy "stereotypy" and uses the symbol S (see Table II). The notation I have used here is that of Hazlett and Bossert (1965) both because it is convenient for behavioral analysis and because all information is given the common symbol H with appropriate subscript. Table II gives symbols and terms used in this paper, along with others found in the literature.

A problem arises, especially in the physical sciences or mathematics, concerning the relationship between information and entropy, since entropy indicates the amount of uncertainty associated with a state of a system, provided categorization is of the finest possible grain (Quastler, 1958*a*). In most cases, many physically distinguishable states are lumped into equal classes; thus entropy, although an upper limit of information functions, is usually so high an upper limit as to be far from the actual value. I am therefore following Quastler in not equating "information" with "entropy" (see also Brillouin, 1960).

III. INFORMATION THEORY AND BEHAVIOR

A. Methods

The first problem of the behaviorist who wishes to use information theory is to establish categories of behavior. This must be done with careful attention to biological "meaningfulness." The number of categories determines the value of H_{max} and thus sets an upper limit to the information present. Most studies of behavioral interactions involve analysis and description of "fixed action patterns" or "displays" which may be "ritualized," and these have been used as the categories for information theory analysis (Hazlett and Bossert, 1965; Dingle, 1969). It is also possible to utilize a finer-grained system and thus include in the analysis a variety of movements which do not fit the criteria for displays (Altmann, 1965). Further analysis of the data may justify lumping of movements if initial splits prove to have been "ignored" by the animals. The method chosen will depend largely on the intent of the study; for example, if the intent is to study the communicatory function of particular displays, other movements may be omitted.

Once categories have been established, the next step is a behavioral matrix, similar to that given in Table I, from which the various probabilities and values of H can be computed. It is possible, if the data are sufficient, to compute values not only for first order interactions, i.e., for following acts occurring after a given initial act, but also for second and third order interactions (cf. Altmann, 1965). In the case of second order interactions, the probabilities are those of following acts when the preceding *two* acts are known, for example, the probability of an attack following the sequence attack–threat in Table I. Similarly, a third order interaction involves the three preceding acts. Needless to say, these computations require large amounts of data so that the sequences in question occur frequently enough for reasonable estimates of probabilities. The investigator must decide if the effort expended in gathering such quantities of data will be justified by the significance of the results. The use of computers in such situations is virtually mandatory.

If the behavior in question involves the communication of direction and distance information, as occurs in honey bees or trail-laying ants, a different method of computation must be used. In the case of direction, in the absence of any message $H = \log_2 360°$; note that this is the equivalent of H_{max} for 360 equally likely alternatives of 1° each (it would of course also be possible to take finer intervals if the measurements were precise enough). It has also been shown (Shannon and Weaver, 1949) that for a one-dimensional normal (Gaussian) distribution with a standard deviation of σ

$$H = \log_2 \sqrt{2\pi e}\sigma \tag{12}$$

This would be the uncertainty after the message is received by the responding animal, e.g., the worker bee or ant being informed of the direction of a food source by a forager. The reduction in uncertainty or transmission for the angle of orientation (θ) would thus be

$$H_{t(\theta)} = \log_2 360° - \log_2 \sqrt{2\pi e}\sigma_\theta \tag{13}$$

In addition to dealing with direction, one must also compute the transmission for distance (d). Thus

$$H_{t(d)} = \log_2 d - \log_2 \sqrt{2\pi e}\sigma_d \tag{14}$$

Then, since we are dealing with both distance and direction, the total amount of information transmitted in the system is $H_t = H_{t(\theta)} + H_{t(d)}$. A more complete discussion of these methods and the rationale behind them can be found in the papers of Haldance and Spurway (1954) and Wilson (1962).

B. Applications

Information theory is a method of description particularly suited to complex social communication. It facilitates analysis of data and can be used to make predictions and to generate models simulating behavior (Altmann, 1965). In addition, it is a convenient tool for the analysis of behavior over time, and in that sense the values obtained are a measure of social "memory." The time involved can either be of short duration, as is the case with higher order behavioral sequences (Altmann, 1965; Hazlett and Bossert, 1965), or of relatively long duration, as in the changes in acts or patterns involved in the establishment of a dominant–subordinate relationship (Dingle, 1969, and this chapter). Finally, in terms of analysis, information theory can be used to estimate the upper limits of communication in a given system or to estimate the minimum amount of behavior necessary to convey certain kinds of information for comparison with the amount actually occurring. This comparison could lead to meaningful evolutionary insights.

A potentially valuable use of information theory is in making comparisons between different kinds of behavior in the same species or between different species or modes of communication. One can ask, for example, if there are differences in the amount of information communicated, between courtship behavior and aggressive behavior, and if such differences as may occur are consistent across species. Similar comparisons could be made between, say, visual and chemical communication both within and between species. There has been some interesting work on structure in relation to function of various pheromones (Wilson and Bossert, 1963; Wilson, 1968) and sounds (Marler, 1959; Marler and Hamilton, 1966), and it would be intriguing indeed if parallel studies were available comparing the structure and function of signals and the amount of information they communicated. A cautionary note, however, must be interjected here. Any valid comparison depends on categories being at least roughly equivalent, so that the values for H_{max} are not unduly biased by whether or not an analysis is fine- or coarse-grained. It would not be valid, for example, to compare H values between species when for one species every movement an animal makes is categorized, while for the other only ritualized displays are considered (see the following for further discussion of the problem of categories).

To conclude this brief discussion of application, I would like to consider Dancoff's principle. As indicated, this principle states that the optimum of redundant information will be that which minimizes the sum of the cost of errors plus cost of redundancy plus cost in information units of error checking, and that, as a result of natural selection, organisms may be expected to approach this optimum. So far there has been no attempt to apply this concept in ethology, although it would seem *a priori* to offer the potential for a powerful analysis of the selective forces acting on behavior. Such an application is not easy and raises questions concerning measurement. The computation of redundancy is not particularly difficult, but how does one measure the cost of errors or of error checking, from the point of view of both natural selection and information theory? In a sense, this raises the fundamental problem concerning the adaptive advantage of various behaviors. As Marler and Hamilton (1966), among others, point out, there is much discussion in the literature of presumed adaptive advantage but little hard data demonstrating its existence. Likewise, such data are difficult to obtain. Nevertheless, these considerations are important to our understanding of behavior, and we should not ignore the necessity for more hard data, even though measurements of error cost, error checking, or, for that matter, natural selection are difficult.

C. Limitations

One of the chief problems in using information theory is the establishment of the appropriate categories; this is particularly acute with respect to behavior. The use by ethologists of "fixed action patterns" or "displays" which have become "ritualized" as behavioral units has been based on the underlying assumption that these units were essentially constant in form and pattern. However, this assumption is probably not justified, especially in the case of aggressive behavior (Barlow, 1968; Hazlett, this volume). The fixed action pattern may well be a special case in a spectrum of behavior, and Barlow suggests the substitution of the term "modal action pattern," which indicates the spatiotemporal nature of such patterns. This term also emphasizes the fact that it is often extremely difficult to determine the precise duration or amplitude of patterns and displays. There is a great need for quantification of these behaviors in terms of both time and space, and the work of Hazlett (this volume) and Salmon and Atsaides (1968) is a notable step in this direction. The problems of defining behavioral units spatiotemporally are, of course, present no matter what types of behavior are used to establish categories and are not unique to information theory; they are present in any sort of quantitative analysis. Information measures are relative, not absolute, and the measures associated with a given set of behavioral categories will depend on the behavior and on the researcher who classifies it. The best strategy for the ethologist using information theory would seem to be an initial fine-grained analysis with later lumping of categories if justified by the response patterns (Altmann, 1965).

Once categories have been established, there is still the problem of determining probabilities. Information measures refer to whole ensembles (Quastler, 1958*b*), not to single events, yet it is with a sample of events that the researcher must deal in calculating various values of H. In determining probabilities, p_i is usually taken to be n_i/N where n_i is the number of events in the ith category and N is the total number of events; this was the way p_i was calculated in our hypothetical example in Table I. The probabilities so calculated, however, only estimate the actual p_i's, and, therefore, the values of H are themselves only estimates (Hazlett and Bossert, 1965; Dingle, 1969) which may be biased (Miller, 1955).

A further difficulty is the fact that the various measures of information and communication are based only on what can be observed. Subtler changes in behavior, such as changes in "mood" not reflected in an animal's immediate response to a behavioral act on the part of another, may not be reflected by these measures. Nor do the measures indicate the "meaning" of a communication. Variations in meaning with context and other factors must be determined by other methods (Smith, 1965, 1968). The presence of context importance, although not its precise nature, can, however, be determined by analyzing higher order sequences (Altmann, 1965) or changes in behavior over several minutes (Dingle, 1969). Also it is impossible to interpret types of behavior which may function in "metacommunication," i.e., communication which affects the interpretation of other communication (Bateson, 1955; Altman, 1962). An example might be a display or act which says, in effect, "interpret everything that I do as aggressive." It must also be remembered that measurements of information present and transmission should not be considered actual amounts of information exchanged between animals, but should rather be interpreted as information transmitted to the observer, who is then, hopefully, better able to predict behavior.

Finally, although the amount of information is in general related to its utility, the relation is not necessarily one of simple proportionality, nor is utility always a monotonically increasing function of amount (Quastler, 1958*b*). In biological systems, rare events, although providing a great deal of information when they do occur, may contribute little to the overall amount of information. They may, nevertheless, be extremely important to the organism. With respect to behavior, copulation may be a case in point, since in many animals it takes place only at infrequent intervals. Further, information does not reflect changes in ordering within an ensemble. For example, the p_i's for certain behaviors in one species may be distributed 0.40, 0.30, 0.20, 0.10 and for the same behaviors in another 0.10, 0.20, 0.30, 0.40. The information present in both systems is, of course, precisely the same, although because of the reversed order of the probabilities, the utilities of the behaviors may be quite different.

To sum up, Quastler (1958*b*) has observed that there are four limitations in applying information theory to biology which need particular emphasis. They are, first, that information measures refer to populations of events and not to individual cases; second, that they are relative rather than absolute; third, that the information capacity of a system is often not completely utilized; and, finally, that, although information measures are related to utilities and mechanisms, such relations are not simple. All these cautionary statements apply particularly to behavior and, even though fairly obvious, can easily be overlooked. The chief way to overcome such limitations seems to be the comparative method.

IV. STOMATOPOD AGGRESSIVE BEHAVIOR—AN EXAMPLE OF THE USE OF INFORMATION THEORY IN BEHAVIORAL ANALYSIS

A. Description of the Behavior

Stomatopod crustaceans of the genus *Gonodactylus* usually display vigorous aggressive behavior when two individuals are placed together in a container (Seréne, 1954; Dingle, 1969; Dingle and Caldwell, 1969). The only exceptions to this are interactions between a male and a receptive female, in which case courtship and eventually copulation occur. Aggressive behavior can be classified into several more-or-less distinct acts which with the exception of "meet" and "does nothing" represent behavior patterns generally classified as fixed (or modal) action patterns. The problems of classifying aggressive behavior, as indicated by Barlow (1968), are evident here; there is, therefore, an element of arbitrariness in such a classification. It has nevertheless proved useful in an information theory analysis of *Gonodactylus* behavior. Briefly, the various acts established for *G. bredini* and *G. spinulosus*, the species discussed in this paper, are as follows:

Meet—a "chance" encounter occurring when one animal is recognized as it enters the visual field of the other with no recently preceding overt evidence of recognition.

Approach—a direct deliberate advance toward another animal.

Meral spread—a simultaneous outward spreading of the enlarged meri of the raptorial (second thoracic) appendages.

Lunge—a short rapid forward movement directed to another animal often with the meri in the meral spread position.

Strike—a blow delivered by one individual to another with the dactyls of one or both raptorial appendages.

Chase—rapid pursuit of another individual.

Grasp—seizing of another individual with thoracic appendages 3–5.

Coil—bending over of the body so that the head is positioned directly over the telson.

Uncoil—straightening out following a coil.

Avoid—rapid and abrupt swimming or turning away from another animal.

Does nothing—no discernible behavioral response to an act performed by another animal.

A more detailed description of these acts is presented in Dingle and Caldwell (1969). The strike, since it involves a physical blow, is the most obviously aggressive of the acts. The meral spread has to the observer all the appearances of a threat display. Both of these acts can be delivered while an animal is in the coiled position. A chase usually indicates that one animal has established dominance, and the grasp seems to indicate extreme dominance, although this was not always the case. The classification "meet" is included since it was a convenient way to indicate those situations in which one animal suddenly encountered another as both were swimming about; in such situations, aggressive interactions usually followed. The category "does nothing" was originally included because in hermit crabs this seems to be an aggressive act (Hazlett and Bossert, 1965). This turned out not to be the case in the stomatopods, and so it was not included as an initial act. There seemed to be an indication, however, that as a following act its probability of occurrence in certain situations was altered, and it was thus retained in the interindividual matrices (Tables V and VII).

B. General Methods

The work on the two species of *Gonodactylus* was done at the Bermuda Biological Station, St. George's West, Bermuda, during the months of June and July in 1966 and 1968. Animals were captured in rocks in the vicinity of the station. *G. bredini* was taken mostly in shallow water just below the low tide line in habitats where *G. spinulosus* was not found. The latter seems rather to be primarily a reef-inhabiting species, where it occurs in coral heads and around the bases of adhering sea fans, sponges, and so forth. The two species overlap in the waters around some of the islands at the southeastern entrance to Castle Harbour. The animals lived in water varying in depth from less than a meter to about 6 m in cavities in the aeolianite and coral rubble strewn over the bottom. It was from this habitat that most of the *G. spinulosus* used were taken. In the laboratory, the animals were kept either in running seawater or in seawater changed at least once a day; they were also fed daily.

For observation of aggressive behavior, two adult animals were placed in a fingerbowl of seawater with a sand-covered bottom. The size of the bowl depended on the size of the animals; for large *G. bredini* (30–70 mm overall length) it was 19 cm in diameter, while for the considerably smaller *G. spinulosus* (18–30 mm) and similar-sized *G. bredini* the bowl was 10 cm in diameter. For each species and for the interspecific interactions, 20 homosexual pairs, 10 male–male and 10 female–female, were observed for 1 hr (*G. bredini*) or 40 min (*G. spinulosus* and interspecific). The last 20 min was cut from all interactions involving *G. spinulosus* because it was found that this species virtually ceased being active by that time. (Note: in the remainder or this chapter "observation period" refers either to 1hr or a 40 min pairing of two individuals, "interaction" refers to an exchange of acts, and "act" refers to the performance by a single individual of a single specific behavior pattern.) The animals were size-matched as closely as possible, especially in the case of the interspecific interactions. A detailed quantitative analysis of the results with *G. bredini* is presented in Dingle (1969); this chapter includes data on *G. bredini* only where appropriate for comparison with that from *G. spinulosus*.

To record the events occurring during aggressive interactions, one observer called out a running account while another took appropriate notes. Time was recorded at 2 min intervals to give a reasonable estimate of the durations of interactions and of the intervals between interactions; act durations can be determined only with cinematography (Dingle, 1969). Interactions were initiated either with a meet or more frequently with the direction of a specific aggressive act by one animal against the other and were terminated when aggressive behavior was no longer apparent to the observer. The possible communicative function of nonaggressive behavior was not considered in these analyses. The data for intervals of 10, 10, and 20 min (and in the case of *G. bredini* an additional 20 min) (see Tables III and V) were then combined into matrices from which various values of *H* were calculated. Additional discussion of these methods appears in Hazlett and Bossert (1965) and Dingle (1969).

C. Intraindividual Sequences

The matrix for intraindividual behavioral sequences for the first 10 min of *Gonodactylus spinulosus* encounters is given in Table III. Similar matrices were constructed for the second 10 and final 20 min of *G. spinulosus*

Table III. Frequency Distribution of Intraindividual Two-Act Sequences in First 10 Min of Encounters Between Two *Gonodactylus spinulosus*

	Following act (*B*)								
Initial act (*A*)	Approach	Meral spread	Lunge	Strike	Chase	Grasp	Coil	Totals	p_i
Meet	0	1	2	4	1	1	8	17	0.072
Approach	2	8	9	31	1	2	13	66	0.28
Meral spread	1	0	3	5	2	2	0	13	0.055
Lunge	0	2	2	6	2	0	4	16	0.068
Strike	1	0	1	9	7	4	31	53	0.23
Chase	0	2	0	31	0	1	4	38	0.16
Grasp	0	0	0	3	1	0	3	7	0.030
Coil	1	4	2	10	3	5	0	25	0.11
Totals	5	17	19	99	17	15	63	235	
p_j	0.021	0.072	0.081	0.42	0.072	0.064	0.27		1.00

Table IV. Intraindividual Values of *H* (Bits) and *R* for *G. spinulosus* and *G. bredini*[a]

	G. spinulosus			*G. bredini*		
Intervals	H_B	H_t	R	H_B	H_t	R
I	2.25	0.40	0.200	2.44	0.58	0.132
II	2.02	0.72	0.281	2.47	0.75	0.121
III	2.18	0.63	0.224	2.51	0.87	0.104

[a] The intervals I, II, III refer to the 10, 10, and 20 min divisions, respectively, of an observation period. Data for *G. bredini* from Dingle (1969). $H_{max} = \log_2 7 = 2.81$ bits.

encounters and for these three intervals of encounters involving *G. bredini* (Dingle, 1969). The numbers in the matrix indicate the number of times a given act followed another act performed by the same individual; no attention was paid in this analysis to acts performed by the other individual and their possible influence. Since the primary concern of these studies was to analyze intraindividual aggressive sequences, such apparently nonaggressive acts as "uncoil," "does nothing," and "avoid" were omitted from these matrices. The category "meet" does not occur as a following act because animals were considered to meet only after they had "ignored" each other for at least a short period; this act therefore occurred only at the start of a sequence. The matrices were obtained from the raw data by considering a series of behavioral acts as a set of two-act pieces. Thus the six-act sequence meral spread–approach–strike–coil–chase–grasp was broken down into the combinations meral spread–approach, approach–strike, strike–coil, coil–chase, and chase–grasp. The data were insufficient in number to do higher order information theory analyses.

From the matrices, various values of *H* (information) and *R* (redundancy) for the intraindividual two-act sequences of *G. spinulosus* and *G. bredini* were calculated (See Table IV). Certain differences between the two species are obvious. Information present in the following acts (H_B) and information transmitted per act (H_t) are less in *G. spinulosus*. As a consequence of the lower H_B values, redundancy (*R*) is higher in this species for these intraindividual sequences. Thus in *G. spinulosus*, compared with *G. bredini*, a higher portion of the available possibilities for carrying information remains unutilized. "Error detection" therefore seems to be of greater importance in the *G. spinulosus* system. Also, because less information is transmitted per act (H_t), each act in terms of its influence on the next act by the same individual is apparently less restrictive.

D. Interindividual Sequences

Two-act matrices of aggressive communication between two individuals were constructed for the first 10, second 10, and final 20 min intervals (intervals I, II, and III, respectively) of *G. spinulosus* observation periods;

Table V. Frequency Distribution of Interindividual Two-Act Sequences in First 10 Min of Encounters Between Two *Gonodactylus spinulosus*

Initial act (A)	Following act (B): Approach	Meral spread	Lunge	Strike	Chase	Grasp	Coil	Avoid	Does nothing	Totals	p_i
Meet	0	1	2	5	1	1	11	4	3	28	0.066
Approach	7	6	6	7	0	0	17	37	6	86	0.20
Meral spread	0	1	1	1	0	0	1	7	2	13	0.030
Lunge	1	0	2	0	0	0	4	20	1	28	0.066
Strike	0	0	1	6	2	1	9	55	5	79	0.19
Chase	0	0	0	0	0	0	0	2	0	2	0.0046
Grasp	0	0	0	4	0	1	2	3	0	10	0.023
Coil	6	3	2	16	0	1	12	9	1	50	0.12
Uncoil	0	0	0	0	2	0	0	1	14	17	0.040
Avoid	5	0	3	5	30	0	9	5	57	114	0.27
Totals	19	11	17	44	35	4	65	143	89	427	
p_j	0.044	0.026	0.040	0.10	0.082	0.0093	0.15	0.33	0.21		1.00

the matrix for the first 10 min of interactions is given in Table V. This table (and the similar matrices constructed for the succeeding two intervals) includes all acts performed by either individual of an interacting pair so long as at least one of them was performing acts characteristic of aggressive behavior. For example, if one animal failed to exhibit a recognizable response to an aggressive act by the other, "does nothing" was recorded as a following act. "Uncoil" and "avoid" are also included here because the other individual could respond to them with aggressive acts, and "avoid" could occur as an act following an aggressive act. Except for the fact that the following acts are now behavior patterns in response to indicated initial acts by the *other* member of an interacting pair, the interindividual matrices were constructed in the same way as the intraindividual matrices (Table III). Again, because of insufficient data, sequences of more than two acts were not considered.

From such interindividual matrices the various values of H and R were calculated for the three time intervals. These are given in Table VI

Table VI. Values of H (Bits) and R for Interindividual Interactions of *G. spinulosus* and *G. bredini*[a]

	G. spinulosus				*G. bredini*			
Interval	H_B	H_t	H_t/IN	R	H_B	H_t	H_t/IN	R
I	2.63	0.64	2.48	0.170	2.73	0.63	1.71	0.139
II	2.56	0.79	2.86	0.191	3.00	1.03	2.09	0.054
III	2.54	0.67	2.53	0.199	2.58	0.78	1.94	0.186

[a] Intervals as in Table IV. Data for *G. bredini* from Dingle (1969). $H_{max} = \log_2 9 = 3.17$ bits.

for both *G. spinulosus* and *G. bredini*. Again, as was the case with intraindividual sequences, H_B or information present in the following acts tends to be less in *G. spinulosus* than in *G. bredini*, although the differences in this instance are less marked. The difference between the two species is thus consistent across both intraindividual and interindividual sequences. Similarly, as a consequence of the differences in H_B, the redundancies (R) are greater in *G. spinulosus*, indicating once more that the two-act sequences of this species possess a higher proportion of unutilized possibilities for carrying information. With respect to H_t, information transmitted per act, the values are lower for *G. spinulosus* in time intervals II and III, but the values for the two species are virtually identical in interval I. In both species, H_t is conspicuously higher in interval II than in either the preceding or succeeding interval. This implies that during this interval individual acts are somewhat more restrictive. In *G. bredini*, the H_t value for a fourth 20 min interval was 0.70 bit and thus similar to the values in intervals I and III.

From the values of H_t and knowledge of the average number of acts per interaction between two individuals, it was possible, by multiplying these two figures, to obtain H_t/IN, the average amount of information per interaction per individual. In the case of *G. spinulosus*, the average numbers of acts per interaction for the three time intervals were 3.88, 3.62, and 3.77 acts, respectively, whereas for *G. bredini* these values were 2.72, 2.03, and 2.49 acts, noticeably less. The chief reason for the disparity is in the use of the act "meral spread" by the two animals. In the first place, *G. bredini* makes use of this behavior pattern much more often; for example, in interval I *G. bredini* performed a meral spread as a following act in 64 instances out of 523 ($p_j = 0.122$) (Dingle, 1969), while *G. spinulosus* (Table V) performed it in only 11 instances out of 427 ($p_j = 0.026$). Frequently when a meral spread is performed the other animal may not be "watching" the individual performing it. Since it presumably does not perceive this act under these circumstances, it cannot respond. In these cases, then, where the meral spread was performed in apparent isolation, the number of acts in an "interaction" was one. This occurred more frequently in *G. bredini* because it performed the meral spread more often. Also, the longest sequences of acts in a given interaction occurred in *G. spinulosus*. The greatest number of acts in one interaction recorded for this species was 16, and there were several instances of ten or more acts. In *G. bredini*, on the other hand, the greatest number of acts was nine.

The temporal pattern of information per interaction (H_t/IN) follows that of H_t, namely, a higher value in interval II than in intervals I and III. Again this was true of both species. There is thus no apparent adjustment of information transmitted, in relative magnitude with respect to the time intervals, by considering the interaction rather than the act as the behavioral unit of information transfer. The difference between the two species with respect to total amount of transmission was, however, affected. The markedly greater number of acts per interaction in *G. spinulosus* resulted in higher values of H_t/IN for this species, even though values of H_t were generally lower than in *G. bredini*. *G. spinulosus*, therefore, transmits less information than *G. bredini*, with the exception of interval I, if acts are considered the behavioral units, but more information than *G. bredini* if interactions are the units involved. This implies that for *G. spinulosus* more behavior must take place for appropriate "interpretations" to be made.

Finally, from H_t/IN and estimates of the duration of interactions, the rate of transmission in bits per second was calculated. The minimum duration of interactions, estimated from frames of a cinefilm of the aggressive behavior of *G. bredini* (Dingle *et al.*, 1967), was approximately 0.33 sec. The minimum duration for *G. spinulosus* was assumed to be the same. Notes taken during the observation periods indicated that two animals of either species could remain coiled for about 2 min, so this was taken as the maximum duration of an interaction. Dividing the various values of H_t/IN by these two figures gives a range of transmission rates of 0.21–8.58 bits/sec for *G. spinulosus* and 0.014–6.27 bits/sec for *G. bredini*. By adding these extremes and dividing by two (Hazlett and Bossert, 1965), the range midpoint was obtained for each species; these midpoints were 4.30 and 3.14 bits/sec, respectively. These will be discussed further below.

E. Interspecific Sequences

Matrices similar to those involving interindividual aggressive communication were constructed for interspecific sequences using the three 10, 10, and 20 min intervals. The matrix for the initial 10 min of interactions is given in Table VII. Pairings were made between animals which were size-matched as closely as possible (within 2 mm). This raised the question of the relative maturity of the individuals of each species. Fully grown *G. bredini* may reach 70 mm or more in length, while the largest *G. spinulosus* reaches only about 30–35 mm (Manning, 1969). The result is that when size-matched the *G. bredini* are young animals while the *G. spinulosus* are mature. As a control, several pairings were made on the smallest *G. bredini* available (21–30 mm) and the results compared with those from larger animals reported by Dingle (1969). No differences in behavior were noted, and in fact no differences were observed even in early postlarval stages (Dingle and Caldwell, 1969), although the sample in this latter case consisted of only four individuals. The 21–30 mm *G. bredini* are sexually mature, because two females of 21 mm each were found with fertile eggs. Presumably, no bias was introduced into the observations by the fact that the *G. bredini* involved were not fully grown animals.

Table VII. Frequency Distribution of Interspecific Two-Act Sequences in First 10 Min of Encounters Between *G. spinulosus* and *G. bredini*

	Following act (*B*)										
Initial act (*A*)	Approach	Meral spread	Lunge	Strike	Chase	Grasp	Coil	Avoid	Does Nothing	Totals	p_i
Meet	0	1	3	9	0	1	3	2	0	19	0.043
Approach	6	2	7	14	0	0	12	10	4	55	0.12
Meral spread	1	1	0	5	0	0	7	2	9	25	0.056
Lunge	0	0	3	1	0	0	9	19	4	36	0.081
Strike	5	7	5	69	0	4	21	46	2	159	0.36
Chase	0	0	0	0	0	0	0	0	0	0	0
Grasp	0	0	0	7	0	0	3	5	0	15	0.034
Coil	4	4	2	15	0	0	12	4	3	44	0.099
Uncoil	1	2	2	1	0	0	1	2	17	26	0.058
Avoid	5	0	1	3	7	0	8	4	39	67	0.15
Totals	22	17	23	124	7	5	76	94	78	446	
p_j	0.049	0.038	0.052	0.28	0.016	0.011	0.17	0.21	0.17		1.00

There are two conspicuous differences in behavior between interacting *G. spinulosus* (Table V) and interspecific interactions (Table VIII). These are in the relative frequencies of "strike "and "avoid" as following acts. In the case of *G. spinulosus*, p_j for "strike" was 0.10, while for "avoid" it was 0.33; the p_j's for the interspecific interactions were 0.28 and 0.21, respectively. The figure of 0.21 is intermediate between the values of avoid for *G. spinulosus* (0.33) and *G. bredini* (0.14) (Dingle, 1969). The interspecific value for "strike" (0.28) is higher still than the value of 0.22 found for *G. bredini* alone. Note also that the p_j for "meral spread" is similar to that of *G. spinulosus* rather than *G. bredini*, where it was 0.12 (Dingle, 1969). In intervals II and III, strike frequency was similar in the two species when each was considered alone but remained high in the interspecific encounters. For *G. bredini*, the p_j's for "avoid" increased to 0.24 and 0.25, while they remained essentially unchanged in *G. spinulosus* and increased only slightly in the interspecific case. The p_j's for "meral spread" remained virtually unchanged in all three instances. No conspicuous differences in other acts were noted.

These differences in behavior become particularly important when we consider the number of acts per interaction. The value for interspecific interactions were 3.90, 3.45, and 3.58, respectively, for the three time intervals. These figures are much closer to the values for *G. spinulosus* than to those for *G. bredini*. Presumably, this is in large part because the frequency of meral spreads was low, unlike *G. bredini* but like *G. spinulosus*. Also, like *G. spinulosus* there were longer interactions present.

The H_B values for interspecific interactions (Table VIII) remain essen-

Table VIII. Values of H (Bits) and R for Interspecific Interactions (*G. spinulosus vs. G. bredini*)[a]

Interval	H_B	H_t	H_t/IN	R
I	2.64	0.50	1.95	0.167
II	2.68	0.68	2.35	0.155
III	2.62	0.59	2.11	0.174

[a] Intervals as in Table IV. $H_{max} = \log_2 9 = 3.17$ bits.

tially constant over time and are more similar to those for *G. spinulosus* than for *G. bredini* (Table VI), although somewhat intermediate between the two species. One of the limitations of information theory is nicely illustrated by these H_B values. Note that for interval *I*, H_B is almost precisely the same for *G. spinulosus* and for interspecific interactions (Tables VI and VIII). Yet, as we have seen, there are noticeable differences in behavior, namely, in the p_j's for "strike "and "avoid." This illustrates the fact that information theory does not take into account changes in specific p_j's but only in the overall frequency distribution.

The values of H_t in the interspecific case are lower than for either *G. spinulosus* or *G. bredini* alone, but show the same temporal distribution. That is, the value for interval II is higher than the values for intervals I and III. The fact that this particular pattern occurs in all three situations strongly suggests that whatever it might mean behaviorally, it is a real phenomenon. The information per interaction (H_t/IN) for the interspecific behavior was intermediate between the values of the two species considered alone. Once again the figure for interval II was the highest, reflecting the H_t for this interval since the mean number of acts per interaction was the least. The relative magnitudes of H_t and H_t/IN for the two species considered separately or together thus show the same temporal distribution for the intervals considered.

Finally, the rate of transmission in bits per second was computed for the interspecific interactions. As with the single-species tests, the maximum and minimum durations of interactions were taken to be 2 min (although in this instance such long durations rarely occurred) and 0.33 sec, respectively. Thus the range of transmission rates turned out to be 0.016–7.05 bits/sec with a range midpoint of 3.53 bits/sec. The median once again, however, was about 1.5–2.0 bits/sec and was probably closer to the higher value, as noted also for *G. spinulosus*.

V. GENERAL DISCUSSION

A. Quantitative Analysis of Behavior

Historically, the customary method of ethology has been to describe behavior observed under conditions as natural as possible and to ascribe to it some communicatory function. Unfortunately, however, most of the work has not been quantitative and the occurrence of communication has been assumed rather than proven. It is true that there have been many ethological studies involving the use of models, and these indicate that animals do respond behaviorally to specific stimulus configurations. Thus the importance of stimulus filtering has been demonstrated (Marler, 1961*a*; Marler and Hamilton, 1966). But without statistical analysis of quantitative data, one can never be certain of the pattern of a particular behavior or whether communication is occurring. In fact, it is not difficult to be misled. An example concerning stomatopods is given in Dingle (1969). In interval I of *Gonodactylus bredini* encounters, a coil follows a meral spread or a strike with high frequency. These high frequencies are, however, not significantly different statistically from the overall frequency of coils; thus meral spread and strike are not directive (in a statistical sense) toward a coil. The simple fact of communication has been demonstrated so rarely in the literature on ethology that it seems worth re-emphasizing here the necessity to do so.

Whole areas of ethology, with their concomitant theories, have been built up virtually in the absence of quantitative studies. An example discussed by Barlow (1968) is the concept of the fixed action pattern. In the absence of hard data, these patterns have seemed remarkably constant and stable, but with the advent of recent quantitative studies, these have proved illusory. The dimensions, both spatial and temporal, do seem to cluster about a mode, which has led Barlow to suggest that "modal action pattern" might be a more appropriate term, but we know little about variance or central tendency. Some recent quantitative studies on fixed action patterns are steps in the right direction (Dane *et al.*, 1959; Salmon and Atsaides, 1968; Hazlett, this volume).

Other aspects of behavior, including some of the classical subjects of ethology, also are in need of more quantitative analysis. One example is the frequency of occurrence of certain types of behavior in specific situations, be they spatial or temporal. The frequencies of many acts, for instance, are a function of distance between animals or of size relationships (Hazlett, 1968). Further, specific changes in the frequencies of certain acts occur with time and have been demonstrated to be a function of the establishment of a dominant–subordinate relationship (Dingle, 1969). A second example is the study of the sequential ordering of behavior. Many of the supposedly classic examples of ordered patterns, occurring, say, between the two sexes in courtship, are not as invariant as they once seemed. There have been recent significant steps in applying quantitative methods to such situations (e.g., Wiepkema, 1961; Nelson, 1964), but they have been few. Finally, as Barlow (1968) has suggested, if there is to be a meaningful rapprochement between neurophysiologists and ethologists, there is a great need for more precise quantitative descriptions and models of behavior. These will facilitate the search for the appropriate correlations between behavior and the mechanisms in the central nervous system which control it, although the magnitude and difficulty of this task should not be underestimated. Shaw's (1968) interesting study points out the many similarities between the mechanisms controlling song pattern in the katydid *Pterophylla camellifolia* and the functioning of relatively simple pacemaker systems in both central nervous system and crustacean cardiac ganglion.

The point about quantitative studies seems worth one final comment. Ethology has been led down many a garden path, has barked up many a wrong tree, and has reached many a dead end. These difficulties are, of course, not unique to ethology; however, if ethologists would with greater frequency make use of those quantitative and statistical techniques which are commonly used in other branches of science, many of the difficulties and consequent polemics could be avoided.

B. Information Theory

Information theory is but one of the quantitative tools available to the ethologist who wishes to study animal communications systems. The general limitations discussed previously should be kept in mind during the following discussion. Specifically, it seems worth mentioning again that the studies done so far, especially those on hermit crabs (Hazlett and Bossert, 1965) and stomatopods, deal only with frequencies of behavioral acts, not durations or intensities. For example, a coil was recorded in the same way regardless of whether it lasted for a few seconds or 1–2 min, and so too was a strike, whether it made solid contact, was only a glancing blow, or missed. The relatively small number of observations precluded any sort of time analysis; Hazlett and Bossert (1965) estimate that observation of more than 10,000 acts would be necessary for temporal studies, while their actual numbers of observations ranged from 3000 to 4000 acts, depending on the species of hermit crab. They did not have to contend with the establishment of any sort of dominant–subordinate relationship, which was not the case with stomatopods. For these animals, it was therefore necessary to subdivide the data temporally, thus reducing further the number of observations per matrix. Also, information theory analysis does not itself say anything about particular functions of behavior; in terms of the logical analysis of language and communication, it deals only with syntactics, the formal study of signals as phenomena, rather than with semantics, the meaning of signals, or pragmatics, their significance (Hockett, 1961; Marler, 1961*b*; Smith, 1968). The relation, if any, of information transmission to these latter functions can be determined only by simultaneous use of other methods of analysis.

Such combinations of methods offer the possibility of some interesting insights into behavior. In both species of stomatopod, for instance, and in the interspecific interactions, information transmitted per act (H_t) and per interaction (H_t/IN) was always highest in interval II. A parallel chi-square analysis demonstrated that this was a transition period in the hour-long series of observations for *Gonodactylus bredini*, probably as the result of the recent establishment of a dominant–subordinate relationship which usually occurs at about this time (Dingle, 1969). Likewise, dominance is usually established near the beginning of this interval in *G. spinulosus* and in interspecific contacts, so it seems reasonable to suppose that this is also a transition period in these situations. That being the case, it is interesting that transmission is highest at this time. Perhaps because of the still tenuous nature of the newly established dominant–subordinate relationship, more information is needed to make the appropriate "judgment" as to how to behave. Note that in the interspecific contacts neither species had a preponderance of the "winners," with *G. bredini* "winning" 11 times, *G. spinulosus* seven times, and two "draws."

The greatest value of information theory is the potential, since the units are the same, for interspecific or cross-modal comparisons. It is of interest to compare the results obtained from stomatopods with those from a similar study of hermit crabs by Hazlett and Bossert (1965); the various values of H and R for the several species of hermit crab studied are given in Table IX. The H_B values are generally somewhat higher in hermit crabs, undoubtedly because $H_{\max}$ is also higher as a result of more categories. There were 14–17 categories in the hermit crab studies, with $H_{\max}$ varying from 3.81 bits to 4.09 bits, whereas with nine categories $H_{\max}$ was 3.17 bits for all interindividual stomatopod behavior. Redundancies also tended to be higher for hermit crabs, indicating a greater unused capacity for information, although there was some overlap in range (compare Table VI with Table IX). The potential for complexity thus seems to be realized to a greater extent by stomatopods than by hermit crabs, and, for whatever reason, "error checking" would seem to be of greater importance in the latter.

With respect to H_t, all values for stomatopod interindividual behavior are higher than those for hermit crabs with the one exception that the lowest

Table IX. Values of H (Bits), R, and Transmission Rates (Bits/Second) for the Hermit Crabs Observed by Hazlett and Bossert (1965)

Species	H_B	H_t	H_t/IN	R	Transmission rates
Clibinarius tricolor	3.03	0.35	0.91	0.205	0.7–1.7
C. cubensis	—	0.37	1.09	—	0.6–1.2
Calcinus tibicen	3.05	0.39	1.16	0.199	0.5–1.9
Paguristes grayi	—	0.41	0.99	—	0.4–1.0
Pagurus miamensis	—	0.42	1.11	—	0.8–2.1
P. bonariensis	—	0.44	1.14	—	0.9–4.4
P. marshi	3.29	0.52	1.35	0.177	0.6–2.6
Pylopagurus operculatus	3.12	0.36	0.85	0.237	0.7–3.6

stomatopod value (0.50 for interval I of interspecific behavior; see Table VIII) overlapped the highest value for hermit crabs (0.52 for *Pagurus marshi*). Hazlett and Bossert (1965) suggest that the relatively high H_t of *P. marshi* when compared with other hermit crabs might be the result of the fact that this species is usually covered with commensals and detritus and thus is well camouflaged; the high H_t might be necessary for conspecifics to make a "judgment" of the apparent size, for example, of a behaving animal. The crab with the next highest H_t of 0.44, *Pagurus bonairenis*, is also camouflaged to some extent, so a similar argument may apply. Stomatopods living in cavities in rocks or coral heads are ordinarily hidden from conspecifics. If this argument concerning hermit crabs holds, stomatopods should require a greater information transmission to form the basis for behavioral "judgments." Another factor increasing H_t in stomatopods could be the potential injuriousness of the strike; in this case, faulty "judgments" could be dangerous, and it would presumably be advantageous for sufficient information transmission to take place to avoid them. An interesting question is raised by the fact that H_t was generally lower in the interspecific contests. This might not be as illogical as it at first seems; if the two species could recognize each other as being different species quickly, questions concerning, say, possible sexual receptiveness would not arise, thus making at least some "judgments" easier.

It is also of interest to compare the bits of information transmitted per interaction (H_t/IN) of hermit crabs and stomatopods. The stomatopod values for this parameter are about double those for the hermit crabs, centering at about 2 bits per interation as opposed to 1 bit. This means that for each interaction a hermit crab makes the equivalent of approximately one two-choice decision while a stomatopod makes two. Whether this is a function of the greater rapidity of stomatopod behavior or of some other factor cannot at present be determined. Much more analysis of data from other species will be necessary before we can answer this kind of question.

For the values so far discussed, the only valid comparisons possible are between hermit crabs and stomatopods; there are simply no other data in the literature. Altmann (1965) has done a similar study with rhesus monkeys, but since his analysis was finer-grained and he did not distinguish between intra- and interindividual sequences or between various kinds of behavior such as maternal, courtship, agressive, etc., his data are not directly comparable.

The values for rate of transmission in bits per second obtained for stomatopods can, however, be compared with such rates calculated for the transmission of distance and directional information by bees and ants (Haldane and Spurway, 1954; Wilson, 1962) as well as with hermit crab rates. The rates for stomatopods were 0.021–8.58 bits/sec with a range midpoint of 4.30 for *Gonodactylus spinulosus*, 0.014–6.27 bits/sec with a range midpoint of 3.14 for *G. bredini*, and 0.016–7.05 bits/sec with a range midpoint of 3.53 for the interspecific interactions. As indicated previously, medians probably fell in the neighborhood of 1.5–2.0 bits/sec. The overall range for the eight species of hermit crab was 0.4–4.4 bits/sec (Table IX) with an overall range midpoint of 1.5 (Hazlett and Bossert, 1965). Wilson (1962) has calculated transmission rates for the honey bee dance, based on the study by Haldane and Spurway (1954), as 0.01–2.28 bits/sec and for the pheromonal food source communication of the fire ant as 0.04–1.39 bits/sec. There is considerable overlap in these figures, but it would be interesting to know more about the differences that occur. The upper limit in hermit crabs, for example, is about double that in honey bees and triple that in fire ants. The stomatopod upper limits are one and a half to two times the upper limit of hermit crabs, two and a half to three and a half times that in honey bees, and as much as six times that in the fire ant. The range midpoints in stomatopods are at least double the overall midpoint for hermit crabs. Are these differences the result of different modes of transmission where one might expect a visual system to possess more rapid means of communication than, say, a chemical one? Or are the differences more the result of what is being communicated, directional information in two instances and aggression in the other? Within stomatopods and hermit crabs might the differences observed be in some way a function of habitat? In order to answer these questions, we once again need more comparative information. It is interesting to note that the upper limits in stomatopods extend into the range of one estimate of 6–12 bits/sec for information transmission in human speech (Quastler, 1958*a*).

Because of the relative nature of information measures and the problems inherent in estimating them, there has been some doubt as to the usefulness of information measures in biology (Quastler, 1958*b*). As far as behavior is concerned, it is, of course, but one of the available tools for quantitative analysis; it has the advantage, however, of being applicable to any situation in which an observational matrix is constructed. Its ultimate value will be in its usefulness in elucidating behavior, but for a determination of such usefulness far more comparative data are required. This is a standard refrain in science, yet in this case five studies, or six if the present paper is included, can hardly form the basis for meaningful judgment on the utility of a method. Certainly, information theory would seem to have the potential to yield insights into behavior and perhaps even to pass the ultimate biological test of elucidating adaptive advantage.

ACKNOWLEDGMENTS

This study was supported in part by NSF grant GB-4937 and by the Graduate College, University of Iowa. I thank Drs. Roy L. Caldwell, Brian A. Hazlett, and Joseph P. Hegmann for useful comments on the manuscript, Mrs. Marian Marsolais for the computer programs, and Dr. Caldwell for aid in data collection.

REFERENCES

Altmann, S. A., 1962, A field study of the sociobiology of rhesus monkeys, *Macaca mulatta*, *Ann. N.Y. Acad. Sci.* **102**: 338–435.

Altmann, S. A., 1965, Sociobiology of rhesus monkeys. II. Stochastics of social communication, *J. Theoret. Biol.* **8**: 490–522.

Baerends, G. P., Brouwer, R., and Waterbolk, H. T., 1955, Ethological studies on *Lebistes reticulatus* (Peters): I. An analysis of the male courtship pattern, *Behaviour* **8**: 249–334.

Barlow, G. W., 1968, Ethological units of behavior, *in* "The Central Nervous System and Fish Behavior" (D. Ingle, ed.) pp. 217–232, University of Chicago Press, Chicago.

Bateson, G. 1955, A theory of play and fantasy, *Psychiat. Res. Rep.* **2**: 39–51.

Brillouin, L., 1960, "Science and Information Theory," 2nd ed., Academic Press, New York.

Dancoff, S. M., and Quastler, H., 1953. The information content and error rate of living things, *in* "Information Theory in Biology" (H. Quastler, ed.) pp. 263–273, University of Illinois Press, Urbana.

Dane, B., Walcott, C., and Drury, W. H., 1959, The form and duration of the display actions of the goldeneye (*Bucephala clangula*), *Behaviour* **14**: 265–281.

Dingle, H., 1969, A statistical and information analysis of aggressive communication in the mantis shrimp *Gonodactylus bredini* Manning (Crustacea: Stomatopoda), *Anim. Behav.* **17**: 567–581.

Dingle, H., and Caldwell, R. L., 1969, The aggressive and territorial behaviour of the mantis shrimp *Gonodactylus bredini* Manning (Crustacea: Stomatopoda), *Behaviour* **33**: 115–136.

Dingle, H., Caldwell, R. L., and Schöne, H., 1967, Behavior of the mantis shrimp *Gonodactylus bredini*, Film produced by Institut für den wissenschaftlichen Film, Göttingen, Germany (Encyclopedia Cinematographica.)

Garner, W. R., 1962, "Uncertainty and Structure and Its Psychological Concepts," Wiley, New York.

Haldane, J. B. S., and Spurway, H., 1954, A statistical analysis of communication in *Apis mellifera* and a comparison with communication in other animals, *Insectes Sociaux* **1**: 247–283.

Hazlett, B. A., 1968, Size relationships and aggressive behavior in the hermit crab *Clibinarius vittatus*, *Z. Tierpsychol.* **25**: 608–614.

Hazlett, B. A., and Bossert, W. H., 1965, A statistical analysis of the aggressive communications systems of some hermit crabs, *Anim. Behav.* **13**: 357–373.

Hinde, R. A., 1966, "Animal Behavior," McGraw-Hill, New York.

Hockett, C. F., 1961, Logical considerations in the analysis of animal communication, *in* "Animal Sounds and Communication" (W. E. Lanyon and W. N. Tavolga, eds.) pp. 392–430, AIBS Publication 7, Washington, D.C.

Lloyd, M., Zar, J. H., and Karr, J. R., 1968, On the calculation of information-theoretical measures of diversity, *Am. Midl. Naturalist* **79**: 257–272.

Manning, R. B., 1969, "Stomatopod Crustacea of the Western Atlantic," University of Miami Press, Miami, Florida.

Marler, P., 1959, Developments in the study of animal communication, *in* "Darwin's Biological Work" (P. R. Bell, ed.) pp. 150–206, Cambridge University Press, London.

Marler, P., 1961*a*, The filtering of external stimuli during instinctive behaviour, *in* "Current Problems of Animal Behaviour" (W. H. Thorpe and O. L. Zangwill, eds.) pp. 150–166, Cambridge University Press, London.

Marler, P., 1961*b*, The logical analysis of animal communication, *J. Theoret. Biol.* **7**: 295–317.

Marler, P. R., and Hamilton, W. J., III, 1966, "Mechanisms of Animal Behavior," Wiley, New York.

Miller, G. A., 1955, Note on the bias of information estimates, *in* "Information Theory in Psychology," Free Press, Glencoe, Ill.

Mittelstaedt, H., 1964, Basic control patterns of orientational homeostasis, *Soc. Exp. Biol. Symp.* **18**: 365–386.

Nelson, K., 1964, The temporal patterning of courtship behaviour in the glandulocaudine fishes (Ostariophysi, Characidae), *Behaviour* **24**: 90–146.

Nelson, K., 1965, After-effects of courtship in the male three-spined stickleback, *Z. Vergleich. Physiol.* **50**: 569–597.

Pielou, E. C., 1966, Shannon's formula as a measure of specific diversity: Its use and misuse, *Am. Naturalist* **100**: 463–465.

Quastler, H., 1958*a*, A primer on information theory, *in* "Symposium on Information Theory in Biology" (H. P. Yockey, ed.) pp., 3–49 Pergamon Press, New York.

Quastler, H., 1958*b*, The domain of information theory in biology, *in* "Symposium on Information Theory in Biology" (H. P. Yockey, ed.) pp. 187–196, Pergamon Press, New York.

Salmon, M., and Atsaides, S., 1968, Visual and acoustical signaling during courtship by fiddler crabs (genus *Uca*), *Am. Zoologist* **8**: 623–640.

Serène, R., 1954, Observations biologiques sur les stomatopodes, *Ann. Inst. Oceanograph. Monaco* **29**: 1–94.

Shannon, C. E., 1948, A mathematical theory of communication, *Bell Syst. Tech. J.* **27**: 379–423, 623–656.

Shannon, C. E., and Weaver, W., 1949, "The Mathematical Theory of Communication," University of Illinois Press, Urbana.

Shaw, K. C., 1968, An analysis of the phonoresponse of males of the true katydid, *Pterophylla camellifolia* (Fabricius) (Orthoptera, Tettigoniidae), *Behaviour* **31**: 203–260.

Smith, W. J., 1965, Message, meaning, and context in ethology, *Am. Naturalist* **99**: 405–409.

Smith, W. J., 1968, Message-meaning analysis, *in* "Animal Communication" (T. A. Sebeok, ed.) pp. 44–60, Indiana University Press, Bloomington.

Stokes, A. W., 1962, Agonistic behaviour among blue tits at a winter feeding station, Behaviour **19**: 118–138.

Wiepkema, P. R. 1961, An ethological analysis of the reproductive behaviour of the bitterling, *Arch. Neerl. Zool.* **14**: 103–199.

Wilson, E. O., 1962, Chemical communication among workers of the fire ant *Solenopsis saevissima* (Fr. Smith). 2. An information analysis of the odour trail, *Anim. Behav.* **10**: 148–158.

Wilson, E. O., 1968, Chemical systems, *in* "Animal Communication" (T. A. Sebeok, ed.) pp. 75–102, Indiana University Press, Bloomington.

Wilson, E. O., and Bossert, W. H., 1963, Chemical communication among animals, *Recent Progr. Hormone Res.* **19**: 673–716.

Chapter 5

PREDATORY BEHAVIOR OF A SHELL-BORING MURICID GASTROPOD

Melbourne R. Carriker and Dirk Van Zandt

Systematics–Ecology Program
Marine Biological Laboratory
Woods Hole, Massachusetts

I. INTRODUCTION

Invertebrate animals in a wide range of taxa have evolved complex mechanisms for excavating boreholes through the mineralized exoskeletons of live prey to obtain food. Although the mechanism of penetration has been the subject of increasing research in recent years (Carriker *et al.*, 1969), predatory behavior has received less attention (Carriker and Smith, 1969).

One of the better-known calcibiocavites is the marine carnivorous gastropod *Urosalpinx cinerea* (Say). This species is of behavioral interest because of the intricate sequence which characterizes detection, approach to, and penetration of prey. This progression may be characterized as follows: (a) distance detection of prey, (b) approach to prey, (c) close-range detection of prey, (d) selection of borehole site, (e) penetration of shell, and (f) feeding through the borehole. Its small size, abunance, ease of maintenance and handling in the laboratory, and relatively short life cycle make this predator an especially useful experimental animal.

In this chapter we review existing knowledge of the predatory behavior of muricid gastropods and report the results of an analysis of the behavior of portions of the predatory sequence of *U. cinerea*. *U. c. cinerea* and the closely related subspecies *U. c. follyensis* Baker (Family Muricidae, Suborder Stenoglossa, Subclass Prosobranchia) were employed as experimental animals.

The research of the senior author was initiated in New Jersey in the summer of 1941 (Carriker, 1943, 1955*a*), was reactivated in the summers of 1956–1958 at the University of North Carolina Institute of Fisheries Research (Carriker, 1957), and has been continued with the junior author at the Marine Biological Laboratory, Woods Hole, Massachusetts since 1965.

II. EXPERIMENTAL ANIMALS

A. Predator

U. cinerea is a slow-moving shallow-water marine snail common on hard surfaces in relatively clean coastal waters (Fig. 1). The species is a major predator of commercial oysters along the East Coast of North America and was transported inadvertently to the West Coast of North America and to Britain in the latter part of the last century (Carriker, 1955*a*). Its distribution is limited to north temperate latitudes. Other shell-boring muricid species replace it in the tropics and in the Southern Hemisphere.

Individuals of *U. c. cinerea* (Say) grow to an average maximum shell height (i.e., length of shell) of 30–33 mm from Delaware Bay northward and

Fig. 1. View of left side of *U. c. follyensis* underwater. Ventral surface of foot is visible through glass. Proboscis tip is extended down propodial indentation, exploring glass–shell juncture. Fleshy siphon tip is extended normally beyond the shell siphon. Shell height of snail 40 mm. Light photograph.

from Chesapeake Bay southward in lower estuaries and along the coast (Myers, 1965). Discrete populations may vary considerably in size, color, and shell shape (Wood, 1965).

U. c. follyensis is a large subspecies whose distribution appears to be restricted to the high-salinity embayments along the eastern shore of Maryland and Virginia. The subspecies was first described by Baker (1951) and was later shown to differ genotypically from *U. c. cinerea* by Blake (1966). Individuals of *U. c. follyensis* reach an average maximum height of 47–50 mm (Myers, 1965). The largest specimen recorded in the literature measured 61 mm in shell height (Galtsoff *et al.*, 1937).

In the Morehead City area, North Carolina, we collected adult specimens of *U. c. cinerea* by hand at low tide on rock jetties at Shark Shoals and on piling in Bogue Sound. Salinities in these embayments during the summers of 1957–1958 fluctuated between 28 and 36 ‰. Snails were feeding principally on oysters, barnacles, and mussles. In Woods Hole, Cape Cod, Massachusetts, snails were collected on rock jetties and large boulders, where they were feeding principally on barnacles. Salinities off Woods Hole fluctuated between 31 and 32 ‰ throughout the year. *U. c. follyensis* specimens were collected for us in Hog Island Bay, Virginia, where they were subsisting primarily on oysters, and shipped airmail to Morehead City during the summers of 1957–1958 and to Woods Hole after 1965. Snails were mailed in small quantities of seawater within plastic refigerator bags packed loosely in soft excelsior over ice in plastic bags to keep them cool. Survival of snails shipped in this manner, even during the warmest months of the year, was close to 100 percent.

Snails were maintained in the laboratory in rapidly flowing seawater in a nontoxic system with an excess of oysters. At the Institute of Fisheries Research, seawater was pumped continuously from Bogue Sound, salinity of the seawater ranged between 31 and 36 ‰ and the temperature fluctuated between 22 and 31°C during the period of behavioral observations. At the Marine Biological Laboratory, water was pumped periodically from Great Harbor and fed into the laboratory from a large reservoir on the roof. Snails were employed in behavioral studies after acclimatization for several weeks to conditions of the laboratory running seawater. During the major periods of observations, July through October, the salinity of the seawater fluctuated between 31 and 32 ‰, the pH between 8.1 and 8.2, and the temperature between 17 and 23°C. For ancillary experiments conducted during winter months, running seawater was warmed by a heat exchanger and then degassed by aeration. Such experiments were repeated in the summer for confirmation, as we do not yet know the physiological effects of keeping snails from hibernating or bringing them out of hibernation. Populations of *U. c. cinerea* were employed in the early descriptive work, and *U. c. follyensis*, because of its larger size, for the later quantitative investigations.

B. Prey

Oysters, *Crassostrea virginica* (Gmelin), were used for food and as experimental prey. In the Morehead City area these bivalves were abundant, forming dense clusters in the intertidal zone. On Cape Cod the species was scarce and was grown in commercial quantities only in a few protected embayments. Large quantities are imported from southern waters and held for market in local estuaries and embayments. We employed both local and imported oysters and maintained them in metal baskets off piers at the two laboratories to provide fast-growing animals attractive to the snails.

III. DETECTION AND APPROACH TO PREY

A. Distance Detection of Prey

The aqueous environment of *U. cinerea* is enriched by an ecologically important, but relatively unexplored, qualitatively and quantitatively fluctuating solution of external metabolites released by living organisms. Some of these chemicals serve as signals which provide critical sensory information and mediate the interactions of *U. cinerea* and its prey. The significance of further information on the chemical nature of these metabolites and their role in behavior, especially relative to control and the possible masking and other effects of pollutants, is obvious. The following section is a review of knowledge in this field relative to *U. cinerea*.

The food of *U. cinerea* consists of a variety of animal species. On the East Coast of the United States the diet includes principally several species of barnacles, small oysters (*Crassostrea virginica*), and blue mussels (*Mytilus edulis*). Live prey are preferred over dead tissues (Carriker, 1955*a*; Galtsoff, 1964; Wood, 1968).

A number of investigators have demonstrated under a variety of conditions in the field that *U. cinerea* can identify live prey up the current as much as several hundred feet (Carriker, 1955*a*). At what distance a snail can select a single type of prey out of a number of available species is not known. Snails can similarly identify prey in aquaria of a wide range of sizes and turbulence of seawater, as well as in controlled conditions in olfactometers (Blake, 1962; Carriker, 1955*a*; Haskin, 1940, 1950; Wood, 1965*a*). This ability suggests that distance detection of prey is solely by chemoreception.

Blake supported the hypothesis of Sizer (1936) and Haskin (1950) that metabolic end-products of prey released in direct proportion to oxygen consumed are the chemical attractants to which *U. cinerea* responds. Wood (1965*a*) confirmed the interrelationships of attractance, metabolic rate, and food intake by prey suggested by Blake and by Janowitz (cited by Blake,

1960). These observations explain why chemotactic orientation to young oysters is more pronounced than to older oysters (Haskin, 1940, 1950); why confinement of prey in aquaria, where they receive less food than in the field, reduces their attractiveness (Blake, 1961; Carriker, unpublished; Stauber, 1943; Wood, 1968); and why snails creep around freshly killed oysters to reach live ones (Federighi, 1931). Wood (1965, 1968) in olfactometric studies concluded that preference of *U. cinerea* for specific prey species is not genetically fixed but depends upon cozonation of prey and predator, relative abundance of prey, and the recent ingestive experience of the predator.

Although an index of attractiveness of prey has been established, the basic stimulus has been identified only as one or more of the metabolic end-products of prey. Modification by Blake (1962) of attractant from the effluent of oysters revealed at least two fractions, a volatile one and one with characteristics of a small protein or peptide. Chemical analyses of oyster metabolites demonstrated the presence of ammonia, urea, and 11 amino acids. In biological assays *U. cinerea* was strongly attracted to ammonia of nonbiological origin and to the volatile fraction from the effluent of oysters, suggesting the identity of the two chemicals. Blake suggested that the volatile fraction stimulates the initial movement and preliminary orientation of these snails to oysters. Wood (1965*a*) in a quantitative analysis of the effluents of several prey hypothesized that the ammonium ion is a nonspecific attractant to which unconditioned *U. cinerea* may respond and that specific amino acids, most likely glycine and taurine, may identify the effluents of *Balanus* and *Crassostrea*, respectively. Later, however, Wood (1966) and Webb and Wood (1967) concluded that patterns of free amino acids in the water are so similar to those excreted by organisms that it is unlikely patterns of prey origin could serve as information-carrying stimuli. Clearly, more research must be undertaken before specific distance chemical stimuli can be characterized.

Experimentation on other carnivorous Neogastropoda and on Mesogastropoda (Kohn, 1961) suggests that in *U. cinerea* the osphradium may play a primary role in distance reception (see also Alexander, 1970). This organ lies in the mantle cavity at the base of the siphon and is bathed by a continuous stream of inhalant seawater from the environment. The part played in distance chemoreception by the mantle edge, tentacles, and propodium is not presently known and could be determined by neural ablation.

B. Approach to Prey

In the field *U. cinerea* individuals living among prey move little, whereas on firm bottom devoid of prey they creep about 5–7 m per day into the current in the direction of food organisms. At current velocities above 0.2 cm/sec snails exhibit a positive rheotaxis and move into the flow, the siphon pointing upstream and the spire of the shell downstream. Friction of the bottom reduces current velocities close to the bottom to a fraction of those higher in the water column. Snails are thus generally not exposed to strong currents in their native microhabitats, and the speed of these flows normally does not affect the rate of creeping (Carriker, 1955*a*).

For short periods, especially after being handled, snails may creep more rapidly than the rates cited. In the laboratory this rate for *U. cinerea* ranges maximally from 2.5 to 2.8 cm/min. Wood (personal communication) observed that *U. c. follyensis* in an olfactometer crept slightly in excess of 3 cm/min. Soft muddy and unstable sandy bottoms devoid of hard objects are unfavorable for locomotion, and these gastropods occur there in insignificant quantities (Carriker, 1955*a*). Irregularities of the bottom establish eddy currents which may promote aberrations in the response of snails to stimuli in the seawater. No one has reported the behavior of *U. cinerea* as it creeps toward prey in the field. Such observations could be carried out by scuba divers and would add important information on the behavioral ecology of the species.

Activities of *U. cinerea* decelerate as temperature of the water drops in the fall. In the range of 7–15°C, varying with latitude and other environmental factors, boring and feeding cease. Between 2 and 10°C locomotion stops, and individuals hibernate, burrowing shallowly in the sediment with siphon tips in contact with the water. As salinities fall below approximately 15 ‰, movements likewise slow down and then stop (Carriker, 1955*a,b*; Hanks, 1957; Manzi, 1970). The effect of lowered salinities on the attractiveness of prey is not known. As the temperature of the water rises approximately above 10°C in the spring, snails exhibit a pronounced negative geotaxis, creeping upward on submerged objects. This response is especially marked in females during the breeding season. The response persists in the dark. In strong light on clear days, snails move away from the source of light under objects; in dim light, they move out into the open toward it; and at weaker intensities, phototactic response is lost (Carriker, 1955*a,b*).

In the laboratory some 50–80 percent of a population of *U. cinerea* will respond to freshly introduced live prey (Carriker, 1955*a*, 1957; Blake, 1960, 1961). The remainder scatter about; some climb the walls of the container, others remain inactive. Snails most responsive to the effluents of freshly introduced prey are those taken from prey in the act of penetrating the shell. The path of active snails toward prey is initially an irregular spiral with many turns. After a choice is made, snails follow a less winding path, slowly rotating the shell from side to side through an arc of about 45°. Due to restriction of the fleshy siphonal canal by the shell, a turning movement of the shell is necessary for the snail to obtain samples of attractant along its path (Blake,

1960). Klinotaxis is less pronounced in some individuals than in others and is more noticeable when a snail creeps out of the direct flow of the attractant. Wood (personal communication) reports that snails also used the propodium to compare effluents in his olfactometric studies. Whether a snail is guided chemically by a gradient of the signal resulting from dilution in running seawater or from attenuation of the potency of the stimulus is not known. Possibly both occur. *U. cinerea,* for example, locates its prey in both recirculated and in continuously running seawater. It is probable that metabolites of prey would become uniformly distributed in recirculating seawater and snails would have difficulty locating prey unless, as suggested by Blake (1962) and Wood (1965*a*), the attractant possesses a relatively short life which results in a gradient of concentration away from the prey detectable by the snail.

C. Close-Range Detection and Mounting of Prey

1. Problems

If we may hypothesize that a chemical like ammonia is a general nonspecific distance attractant which stimulates the initial movement and preliminary orientation of *U. cinerea,* then how is close-range chemical identification of prey made? Blake (1962) suggested that this might be done by detection of specific protein or peptide fractions in the effluent. Janowitz (cited by Blake, 1960) observed that *U. cinerea* was attracted to oxaloacetic acid and might be able to detect it near rapidly growing or metabolizing oysters, and that oysters fed on *Cryptomonas* grew rapidly and were attractive, whereas oysters fed on *Skeletonema* grew more slowly and were less attractive. Wood (1965) showed that the attractiveness of barnacles was higher when the barnacles were fed *Artemia* larvae than when the barnacles were fed only algae. Stauber (1943) suggested that the snail may confirm the immediate presence of its prey by creeping to the excurrent siphon. Our observations, however, demonstrate that the snail may mount prey on any side. It is evident that we know little about the nature of the chemical signal(s) involved in detection of prey, including whether distance and close-range cues are the same, or different, chemicals.

As indicated by its behavior, *U. cinerea,* once in the immediate vicinity and in contact with prey, employs receptors on the propodium to confirm the presence of prey. When a short distance from active prey, or when prey are introduced close by, snails often raise the anterior part of the foot and, "standing" on the posterior tip of the foot, propodium and tentacles fully extended, swing the propodium back and forth in a pattern suggestive of searching. The full dilation of the tentacles indicates that these appendages may also be involved in prey recognition, but this has not been verified. Kohn (1961) described similar responses in other predatory gastropods.

These observations suggest that in identification by *U. cinerea* of the immediate presence of prey, distance–close-range attractant(s) may be reinforced by attractant absorbed to the exterior of the shell of the prey, by chemicals released by the eroding surface of the shell of prey, by the opening–closing movements of the valves of prey, and/or by the exterior topography of the shell of prey. These possibilities were examined in Experiments (a) to (k) carried out with *U. c. follyensis* during the warm months of the period 1966–1970 at the Marine Biological Laboratory. The response of snails to live oysters and to single oyster valves treated in a variety of ways was measured by the rate at which snails mounted and bored holes in the valves.

2. Experimental Laboratory Studies

In each of the following experiments (a) to (i), the following procedure was employed. Observations were made in a clean, shallow fiberglass tray 71 cm wide and 130 cm long, in seawater 6.5 cm deep. Flourescent illumination, at ceiling height, was uniform over the tray and was on during the day. The range of temperature of the seawater during the experiments was 19–21°C, salinity was 31–32‰ and flow was 5–6 liters/min. Seawater was introduced at one side of the tray and ran continuously. Tests with methylene blue indicated that the water flowed circularly around the rectangle, moving maximally near the periphery and decreasing inward toward the center. It drained over a standpipe at one end of the tray. For each experiment, 24 oysters, ranging in length from 55 to 95 mm and scraped and brushed clean of barnacles, young oysters, and other organisms, were numbered consecutively with india ink and arranged in an oval in the tray just inside the path of the most rapidly flowing seawater (Fig. 2). The oval arrangement, rather than a circular one, was employed to conform with the rectangular shape of the seawater tray. For each experiment, a total of 120 *U. c. follyensis* specimens, ranging in height from 25 to 40 mm and deprived of food for 2

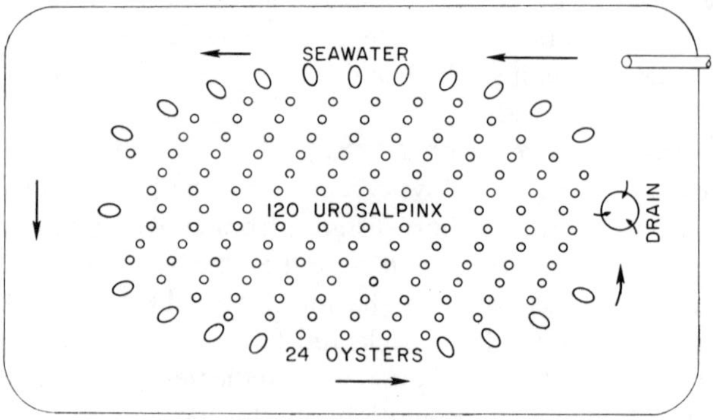

Fig. 2. Initial arrangement of snails and oysters in experimental seawater tray.

weeks, were placed randomly inside the oval of oysters. From this position, active snails crept toward the oysters, stimulated by the current and by attractant from pumping oysters. No attempt was made to distinguish the crawling behavior of males and females. In the absence of live oysters, as well as in the presence of less attractive oysters (that is, those metabolizing less actively), a small proportion of the snails responded to metabolites in the inflowing seawater from organisms fouling the seawater pipes. Each day between 0800–0900 hr we counted the number of snails on each oyster and returned snails outside of the oval of oysters, including those on the sides of the tray, to the area within the oval. This was done to increase potential contacts of snails with oysters. We did not disturb snails on oysters. At each daily observation we also noted which oysters were gaping and which were partly or completely consumed (Table I). At the end of the 7-day period we terminated the experiment and counted the number of boreholes. Inconspicuous breaks and points of incomplete closure at valve edges not detected in cleaning were revealed by immersion of intact oysters in 1 percent methylene blue in seawater for about an hour before they were opened. The dye penetrates at these sites and stains the tissues inside. All oysters were then opened. All oysters which had not been perforated by snails, including those whose valves were bound, were in good condition.

Because it was impossible without lifting snails from oysters to determine whether they were boring and because nonboring snails moved on and off of oysters at an unknown rate, the number of snails recorded daily on oysters provided only a general index of attractiveness to oysters. Borehole excavation, on the other hand, represented clear recognition by snails of live prey and provided an easily quantitated index of response. Snails mounted oysters at successive times during the 7-day period. This accounted for the fact that boreholes ranging from completed ones to those just begun were observed at the end of experiments.

a,b. Response of *Urosalpinx* to (b) Bound Oysters in Proximity of (a) Unbound Oysters. Experiments (a) and (b) were designed to determine to what extent snails, attracted by oyster metabolites, will mount and bore nonpumping oysters in the vicinity of normally pumping oysters. Each of 12 live oysters, after the valves had been smoothed slightly with a file, was bound tightly with two to three turns of stout rubber bands $\frac{1}{4}$ inch wide to prevent opening and closing of the valves and discharge of attractants. Twelve other oysters were left unbound to release attractants through normal pumping. Bound oysters were alternated with unbound ones in the seawater tray. Rubber bands were leached in seawater, and snails did not respond to them.

Snails bored all and consumed most of the unbound oysters in the week (Table I). By the third day, two oysters were drilled and consumed; by the

Table I. Response of *U. c. follyensis* to Bound Oysters (*C. virginica*) in Proximity to Unbound Oysters (Experiments a and b)

Oyster No.	Number of snails on each oyster at daily observation								Number of boreholes per oyster	
	Day 1	Day 2	Day 3	Day 4	Day 5	Day 6	Day 7		Incomplete boreholes	Complete boreholes
					a. Unbound oysters					
1	3	2	3	7	14[a]	2	1		0	2
3	3	3	4	3	4	5	5		1	2
5	5	4	5	9	0[a]	0	0		1	2
7	4	5	5	7	18[b]	0[a]	1		3	3
9	4	4	6	10[b]	0[a]	0	0		1	3
11	2	3	3	2	4	17[b]	0[a]		1	2
13	2	8	13[b]	0[a]	0	0	0		1	2
15	6	2	3	7	8	2[a]	1		2	2
17	5	7	0[a]	0	0	0	0		4	1
19	5	4	3	4	5	17[b]	25[a]		2	1
21	6	9	0[a]	0	0	1	1		3	1
23	4	3	3	8[b]	0[a]	0	0		1	3
Total	49	54	48	57	53	44	34	339	20	24
					b. Bound oysters					
2	1	1	0	0	1	0	0		0	0
4	0	0	0	1	0	1	1		0	0
6	2	0	0	1	3	2	5		2[c]	0
8	0	0	0	1	0	0	2		0	0
10	1	0	3	1	0	0	5		2	0
12	6	4	6	8	10	4	6		3[c]	1[c]
14	3	1	2	2	0	0	0		1	0
16	2	2	3	4	6	2	1		1	1
18	2	4	0	0	0	2	5		2	0
20	3	2	2	8	5	5	1		1[c]	1
22	1	0	0	0	1	1	1		0	0
24	1	1	0	0	0	2	0		0	0
Total	22	15	16	26	26	19	27	151	12	4

[a] Oyster completely consumed.
[b] Oyster gaping.
[c] At edge of valve.

fourth day, one; by the fifth day, four; by the sixth day, two; and by the seventh day, two. As soon as the flesh was consumed, snails abandoned the valves. The number of snails on bound oysters was slightly less than half that on unbound oysters, and the number of boreholes on bound oysters was approximately one third that on unbound ones. There were three times as many incomplete as complete boreholes in the bound oysters, compared to about equal numbers of each in the unbound oysters. This resulted from the fact that snails were slower to initiate boring into the bound oysters.

Examination of the 12 bound oysters disclosed that edges of the valves of seven of them had one or more minute openings which permitted diffusion of fluid from the mantle cavity. So far as we could tell, it was only these oysters which were bored. The total low number (16) of boreholes in the bound oysters demonstrated that binding, although imperfectly sealing the mantle cavity of seven of them, did significantly reduce their attractiveness even in the close presence of pumping unbound oysters. Attractant from pumping oysters probably did not stimulate snails to bore bound oysters. This is suggested by the low number of snails on, and the absence of boreholes in, bound oysters 2, 4, 8, 22, and 24, from which so far as we could determine there was no release of attractant (Table I). Attractant adsorbed to the surface of bound oysters may have been eclipsed by attractant freshly released in the water by pumping oysters. This probability finds support in stock laboratory populations of snails in which oysters which do not open their valves are generally not bored.

Unexpected, however, was the discovery that about half of the boreholes in bound oysters occurred at valve edges at or near points where fluid diffused from the mantle cavity, whereas in the unbound oysters all boreholes were excavated away from the edges. This finding suggests that valvular movements of prey inhibit boring at the juncture of the valves and that, as soon as these movements are blocked, boring takes place at points where attractant is escaping.

Snails on unbound oysters, in response to fluids draining from oyster flesh injured by the feeding predator, increased rapidly in number with gaping of perforated valves and then decreased sharply after the flesh was removed. The staggering of gaping during the week accounted for the generally uniform number of snails on oysters per day.

c. Response of *Urosalpinx* to Bound Oysters. Experiment (c) was run to determine whether snails will mount and bore bound oysters in the absence of attractant from normally pumping oysters. Twenty-four oysters were bound with rubber bands in the manner described in experiments (a) and (b) and arranged in an oval in the tray with snails.

In spite of careful examination at the time of cleaning and binding, we found that 18 of the 24 oysters were incompletely closed by the rubber bands and these were bored (Tables IIA and IIB). The remaining six oysters were tightly closed and escaped penetration. In the absence of unbound oysters, the bound oysters attracted a total of as many snails as did the unbound oysters in experiment (a). This response suggests that attractant from pumping oysters did partially shield the bound oysters from predation in experiments (a) and (b) and that in the absence of this buffer in experiment (c) attractant from minute openings in the valve edges of the bound oysters attracted proportionately more snails. These suggestions are supported by the fact that the number of snails on imperfectly sealed oysters in experiment (c) increased daily, whereas there was not clear-cut increase on imperfectly sealed oysters in experiment (b). Twelve of the boreholes in experiment (c) were completed. This confirms the similar occurrence in experiment (b) and demonstrates that the stimuli of valvular movements and normal pumping by oysters are not required for completion of the excavation. Seventeen of the 30 boreholes were excavated at valve edges at points where the closure was incomplete. This demonstrates, as it did in experiment (b), that recognition of an apparently minute amount of attractant is sufficient stimulus for the snail to initiate and complete shell penetration.

Seeking to seal oysters reliably in further experiments, we resorted, after unsuccessful testing of a variety of adhesives, to sealing with plaster of paris and coating the plaster with Dekophane (Rone Pearl Corp.). Because of the moribund condition of oysters after sealing with these materials for a week and the slightly repellent nature of the Dekophane, we abandoned the attempt until a more functional sealer can be found. An effective method of sealing oysters without affecting their health would be helpful in continued studies of the response of these snails to oyster attractant.

d. Persistence of Boring by *Urosalpinx* when Oysters are Bound After Initiation of Boring. In experiments (b) and (c), snails bored prey successfully in the complete absence of valvular movements. This raised the question of whether snails can continue boring if valvular movements of prey are interrupted after initiation of boring. The matter was tested in experiment (d).

Twenty-six unbound oysters were placed with snails in the experimental seawater tray. On day 1 we removed the six oysters which had not been mounted by snails. On day 3 we gently bound the remaining 20 oysters with rubber bands and in the process knocked off eight snails. Six of these resumed boring when replaced over their incomplete boreholes. We left the remaining 18 oysters with 32 snails on them in the tray for another 4 days. During this period only three snails abandoned boreholes.

At the end of the week when we terminated the experiment, we counted a total of 24 boreholes in the valves (Table IIB). Eight of the snails on oysters

Table IIA. Summary of Responses of *U. c. follyensis* to Live Oysters (*C. virginica*) and Oyster Valves Treated in a Variety of Ways to Reduce or Eliminate Attractant

Experiment	Condition of oysters and empty valves	Total number of oysters or valves in experiment	Total number of snails in experiment	Number of snails on all oysters or valves each day						
				Day 1	Day 2	Day 3	Day 4	Day 5	Day 6	Day 7
a	Unbound live oysters	12	120	49	54	48	57	53	44	34
b	Bound live oysters	12		22	15	16	26	26	19	27
c	Bound live oysters	24	120	32	42	48	53	57	56	62
d	Live oysters bound third day	18	32	—	—	—	32	32	28	28
e	Bound alcoholized live oysters	24	120	11	11	16	22	23	34	48
f	Fresh bound oyster valves	24	120	4	12	24	14	17	11	20
g	Heated bound oyster valves	24	120	1	2	3	9	7	8	7
h	Fresh single oyster valves	24	120	3	6	3	4	4	3	4
i	Heated single oyster valves	24		1	0	1	1	2	4	4

Table IIB. Summary of Responses of *U. c. follyensis* to Live Oysters (*C. virginica*) and Oyster Valves Treated in a Variety of Ways to Reduce or Eliminate Attractant

Experiment	Condition of oysters and empty valves	Total number of snails on all oysters or valves per week	Total number of snails on sides of tray per week	Number of oysters or valves bored	Number of boreholes per week			
					Incomplete	Complete	Total	Total at edge
a	Unbound live oysters	339	—	12	20	24	44	0
b	Bound live oysters	151		7	12	4	16	7
c	Bound live oysters	350	—	18	18	12	30	17
d	Live oysters bound third day	32	—	18	7	17	24	0
e	Bound alcoholized live oysters	165	—	15	13	6	19	13
f	Fresh bound oyster valves	102	412	0	0	0	0	0
g	Heated bound oyster valves	37	420	0	0	0	0	0
h	Fresh single oyster valves	27	569	0	0	0	0	0
i	Heated single oyster valves	13		0	0	0	0	0

had not bored, apparently using the oysters as resting sites. There were no boreholes at the valve edges. This would be expected from the results obtained in experiments (a) to (c), because snails mounted oysters before oysters were bound. The experiment proved that snails can continue boring and complete holes after the oyster ceases valvular movements and pumping of water through the mantle cavity.

e. Response of *Urosalpinx* to Bond Alcoholized Oysters. The purpose of experiment (e) was to test the response of snails to the surface of shell altered by alcohol, a treatment intended to change or remove attractant adsorbed to the shell surface. Conditions of experiment (c) were duplicated except that the 24 oysters were soaked in tap water for 15 min and air-dried, and then the exterior surfaces were swabbed with absolute ethyl alcohol and upended to dry before binding with rubber bands. Bound oysters were left 45 min in the experimental seawater tray in running seawater to rinse before the snails were added.

As suggested by the low total number (165) of snails on the oysters, denaturation decreased the attractiveness of the shell surfaces (Table IIB). The number of snails on the alcoholized oysters did not approach that (350) on bound normal oysters (experiment c), even though there were unanticipated leaks at the valve junctures in 15 of the 24 oysters. A total of 19 boreholes resulted. Thirteen were located at valve edges, confirmation of similar findings in experiments (b) and (c). Nine of the oysters were tightly closed and escaped boring.

Experiments (a) to (e) were concerned with the attractiveness to snails of whole living oysters. The following epxerimenst (f) to (k) were designed to test the possible attraction of valves alone (the live animal removed) under various conditions.

f. Response of *Urosalpinx* to Freshly Cleaned Bound Oyster Valves. In experiment (f) the attraction of snails to the exterior of closed empty valves was tested. Twenty-four oysters were scraped and brushed free of encrustations and popped open at the hinge with an oyster knife, and the flesh was removed. The single valves were washed in tap water, dried, numbered with india ink, dried again, and soaked in running seawater several hours. Matching valves were then bound together with rubber bands, leaving only a slight gap where the oyster knife had penetrated. Bound valves were then arranged in an oval with snails in the experimental seawater tray and observed daily for a week. As usual, snails which crept outside the oval of bound valves were returned to the center.

As demonstrated by the complete absence of boreholes, snails were not stimulated to bore by the substance of the shell (Table IIB). Snails wandered on and off the bound valves frequently, the number of snails on valves at each daily census increasing slightly with time. The high proportion (two thirds) of snails which crawled out of the oval onto the sides of the tray, especially in the vicinity of the inflowing seawater (one half these snails), indicated the relatively low attraction of the valves and the greater attraction of the inflowing seawater from fouled seawater pipes.

g. Response of *Urosalpinx* to Clean Bound Heated Oyster Valves. The purpose of experiment (g) was to test the attraction of snails to the exterior of empty valves treated with heat. Matched valves bound with rubber bands were prepared in the manner described in experiment (f), except that whole live oysters were boiled for 30 min in tap water, and the flesh was removed without injuring the hinge after the cooked oysters gaped.

No boreholes were excavated in the valves during the week (Table IIB). Furthermore, few snails crept onto the valves, the daily and total number of snails observed on them being significantly less than on the fresh valves in experiment (f). The difference suggests that heating made the valves slightly repellent. This suggestion is supported by the fact that after the third day the number of snails on the valves increased slightly, perhaps because of leaching of the shells by seawater. About the same number of snails crawled out of the oval and onto the sides of the tray as in experiment (f).

h,i. Response of *Urosalpinx* to (h) Freshly Cleaned and to (i) Heated Single Oysters Valves. Experiments (h) and (i) duplicated conditions of experiments (f) and (g), but with single valves, to determine if bound valves are more attractive to snails than single valves. We opened the hinge of 12 live oysters and of 12 cooked oysters (boiled 30 min in tap water) which had been scraped and brushed free of encrustations. Single valves were dried, numbered consecutively, and soaked in running seawater for about two days to remove the mucus and fragments of flesh from the inside. Valves were then placed in a grid pattern in the experimental tray, rows of six fresh valves alternating with rows of six cooked valves. The snails were scattered randomly among the valves, but not on them, and observed daily for 6 days. Snails which crawled onto the sides of the tray were returned among the valves.

No boreholes were excavated (Table IIB), and very few snails were observed on the valves. A calculation at each daily observation of the number of snails on the bottom of the tray and on the valves showed that the average daily density of snails on an area of the bottom equivalent to that of a valve was about seven; on fresh valves, about four; and on cooked valves, about one. This observation confirmed those noted in experiments (f) and (g), that heated shell material is less attractive than fresh shell, and possibly slightly repellent, and that snails unattracted by prey tend to remain on the bottom around the valves or to crawl toward the inflowing seawater. It is significant, however, that four times the number of snails (102) mounted the bound fresh valves in experiment (f) as (27 snails) the fresh single oyster valves in experiment (h), and about three times the number of snails (37) mounted the bound heated valves in experiment (g) as (13 snails) the heated single valves in experiment (i). The explanation for this is not clear. It may simply reflect the tendency of these animals to climb onto higher objects (negative geotactic response) at summer temperatures. Whether the mounting response was also influenced by topography of the bound valves which simulated whole oysters was not ascertained. This possibility should not be dismissed until appropriately tested.

j,k. Response of *Urosalpinx* to (j) Leached Single Oyster Valves and (k) Single Oyster Valves Exposed to Oyster Attractant. Experiments (j) and (k) were designed to test the response of snails to attractant adsorbed experimentally to the shell surface of oysters. This was attempted in experiments (a) and (b), but release of attractant by imperfectly bound oysters confused the results.

Twenty freshly cleaned numbered control oyster valves were placed in running seawater to leach; 20 other clean numbered experimental valves were mixed with 75 live actively growing pumping oysters in a 3 gallon glass tank receiving fresh seawater at the rate of 18 liters/min. The tank was submerged slightly in a large stock tray which contained 880 specimens of *U. c. follyensis*. Valves remained under these conditions for 4 days. During this period 250 snails clustered around the outside of the bottom of the tank of oysters, 225 climbed up the outside surface of the tank, and 231 crawled into the tank among the oysters and valves (Fig. 3). This typical response demonstrated clearly the attractiveness of the oysters.

As quickly as possible (in about 5 min) at the end of the 4-day period, we arranged the 40 valves evenly in the experimental seawater tray in a grid pattern, alternating rows of control and experimental valves. Snails were then scattered uniformly, aperture down, among the valves, but not on or touching any of them. The flow of fresh seawater in the tray was reduced to 2 liters/min to provide a gentle circulation which would not dilute the attractant too rapidly. At first, the number of snails on each valve was counted every 15 min, then at longer intervals (Table III).

A total of 2.5 times more snails mounted the experimental than the control valves (Table III). Whether the significant response of snails to valves exposed to oysters was indeed a result of attractant adsorbed on the shell surface is open to question. For one thing, the initial low level, followed by a sustained high level, of response to the valves is not consistent with the suggested short life of attractant in seawater (Blake, 1961). A similar, though less pronounced, pattern of response was noted on the control valves. There was, furthermore, a similar initial increase followed by a relatively uniform

Fig. 3. *U. c. follyensis* crawling into glass tank in response to attractant from oysters. Snails 35–40 mm high. Light photograph.

number of snails on bound fresh and on bound heated oyster valves in experiments (f) and (g), and on heated single valves in experiment (i). On fresh single valves in experiment (h) there was also a sustained daily number of snails, but no initial increase. The reason for this escapes us.

The unanticipated results of experiments (j) and (k) suggest the hypothesis that snails responded to attractants released by microorganisms on the valve surfaces and that enrichment by the effluent of live oysters during immersion of experimental valves in the tank accelerated the growth of organisms on them and thereby augmented their attractiveness to the snails. This suggests that the increase in the number of snails on the bound live oysters in experiments (c) and (e) may have resulted from an increase of distinctive microorganisms on the valves as well as from seepage of metabolites. The hypothesis is supported by Korringa's (1951) report that the rich epifauna (the majority of which originates in the plankton) on oysters differs noticeably from the usual bottom fauna in the vicinity of oysters.

3. Recapitulation

Experiments (a) to (k) amply confirmed the marked attractiveness to *U. c. follyensis* of a factor in the exhalant water of *C. virginica*. So far as we

Table III. Summary of Responses of 120 *U. c. follyensis* to (j) 20 Clean Leached Oyster Valves and to (k) 20 Oyster Valves Exposed to Oyster Attractant

		Total number of snails on all valves at each time (hr) of observation																				
Experiment	Valves	0	0.25	0.50	0.75	1.00	1.25	1.50	1.75	2.00	2.25	2.50	2.75	3.00	3.25	3.50	4.00	4.50	5.00	21.00	29.50	Total
j	Leached	0	6	6	7	9	11	10	9	11	12	13	13	11	10	9	13	13	12	9	14	198
k	Attractant	0	8	15	15	22	23	23	25	29	28	28	31	33	36	33	32	31	34	23	25	494

could determine, snails bored only those oysters from which there was release of attractant. The quantity of attractant escaping from incompletely closed bound oysters must have been very small. Tightly closed oysters were not bored. Attractant in the water from pumping oysters was insufficient to stimulate snails to bore tightly closed oysters nearby. Heated shell and shell denatured with alcohol were less attractive to snails than normal shell. Something on the surface of oyster valves, possibly microorganisms attracted to them and enriched by effluent from pumping oysters, attracted snails but did not stimulate them to bore the shell. Valvular movements and normal pumping of seawater by oysters were not required for hole boring. In the absence of valvular activity, many snails bored at the edges of valves at sites where they detected seepage of attractant from the mantle cavity.

Primary recognition of the immediate presence of prey by *U. cinerea* thus appears to depend on identification of a chemical cue in the exhalant water of prey. The extent to which recognition is reinforced by valvular movements, by chemical attractant adsorbed to the shell surface, by distinctive sessile microorganisms on the exterior of the shell, and possibly by topography of the prey was not revealed by these studies. The subject of reinforcement is of sufficient interest, however, that it merits further study. We did demonstrate that none of these four factors is necessary to initiate and to carry boring to completion, attractant from the mantle cavity being sufficient to do this. It is not known, however, whether valvular movement and possibly the rate of release of attractant by the prey accelerate the rate of boring. More sensitive methods of bioassay than were used in our experiments will be necessary to demonstrate whether attractant can be adsorbed experimentally to shell surfaces.

IV. PENETRATION OF PREY

A. Selection of Borehole Site

In the field the right or flat valve of *C. virginica* is often uppermost and is the one which is generally perforated, but in dense clusters or when oysters are resting on their sides boring may occur through either valve (Pope, 1910–1911; Federighi, 1931; Stauber, 1943). In the laboratory we have observed that whichever valve is up is the one which is usually penetrated. Since *U. cinerea* is negatively phototactic to bright illumination, the valve of prey selected for excavation is also probably influenced by the degree of exposure to light. This relationship has not been explored.

In a series of detailed plottings of the distribution of perforations by *U. cinerea* in the shells of *C. virginica* in the field and in the laboratory in the Woods Hole, Massachusetts, area, Pope (1910–1911) discovered that although boreholes are generally distributed over the surface of the shell, the middle areas of the valves, which are the thickest, are somewhat more frequently the site of penetration. He found no evidence to indicate, as had been suggested by earlier observers, that perforations are confined to the limits of the adductor muscle of the oyster or that the snail preferentially selects for attack depressions or the thinnest portion of the shells of its prey. Pope's observations have since been confirmed by Federighi (1931), by Hancock (1959), and by us (Fig. 4). That areas of the valves away from the edges are favored in boring probably reflects the clear-cut avoidance of the valve edges because of valvular motion.

After crawling onto an oyster the snail undertakes a series of exploratory activities leading to selection of the penetration site. The duration of these explorations is highly variable, ranging from individuals which choose a penetration point in a few minutes to those which may crawl for half an hour or so over a valve before initiating boring. Such variables as the amount of attractant and the state of hunger of the snail may influence the length of the exploration. During the search the proboscis is extended intermittently over the midanterior face of the propodium to the shell surface and, its tip undulating with minute wave-like movements, is passed slowly over the substratum, stopping now and then to rasp at live encrustation. The lateral ridges of the propodium may or may not be raised around the protracted proboscis. The proboscis is extremely pliant and flexible, can be moved in any plane, and can be inserted into seemingly inaccessible crevices (Fig. 5). The length of the fully extended proboscis approximates the height of the shell of the snail (Carriker, 1943). What determines the specific site has not been ascertained. Nor is it known whether individuals express consistent

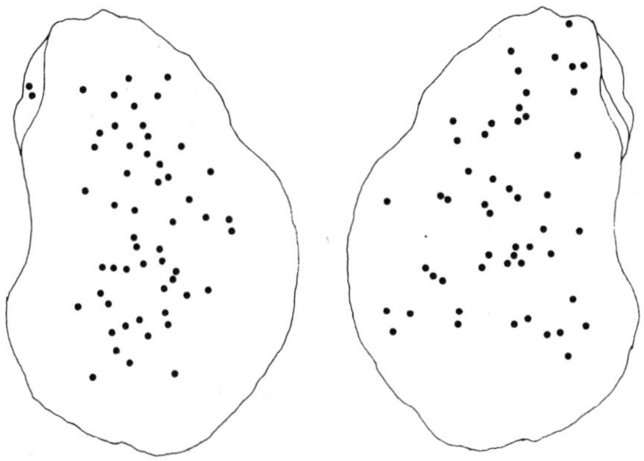

Fig. 4. Distribution of holes bored by 105 *U. c. follyensis* in cupped and flat valves of oysters in seawater tray. Length of valves 8 cm.

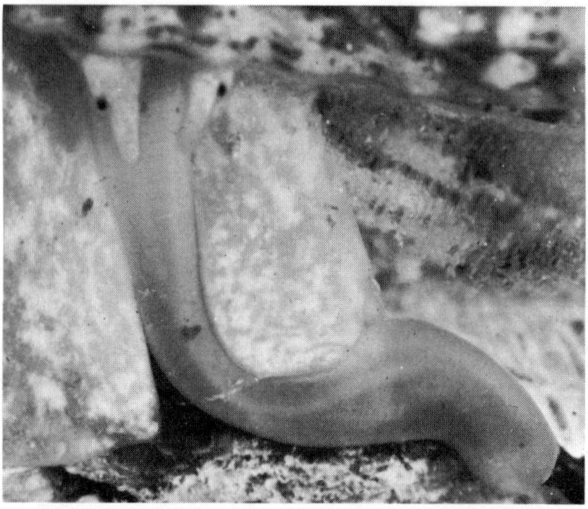

Fig. 5. Anterior view of *U. c. follyensis* under glass showing tentacles and eye spots extending under shell, propodial ridges with everted proboscis passing between them under foot to the shell surface. Inflated distal end of proboscis 1.5 mm in diameter, snail 40 mm high. Light photograph.

preference for a particular portion of the shell surface on successive prey, whether an environmental cue plays a part in selection, or whether selection is random. A break anywhere on the valves of prey through which metabolites diffuse is readily located and used as a penetration site in lieu of boring. A snail will also locate and feed through artificial boreholes drilled experimentally in the valves for them. More than one snail may approach and perforate an oyster simultaneously; in crowded laboratory conditions with a limited supply of food, so many snails will mount an oyster as to completely hide it from view.

The rare reports of snails boring through dead shell (Carriker, 1955*a*; Carriker and Yochelson, 1968; Hancock, 1959) may be explained by the inability of an occasional snail to distinguish clearly living from nonliving prey in close proximity to each other and in the presence of a high concentration of attractant.

Once a point has been selected for penetration, the snail rasps the surface of the shell free of encrustations and loose periostracum and begins the excavation.

B. Penetration of Shell

1. Persistence in Penetration: The Valve Model

The persistence of *U. cinerea* in completing its borehole in the shell of prey is well known. An extreme observation was that by Orton (1930), who described the case of a snail which still returned to its boring position after four interruptions during which it was lifted a few centimeters away from its perforation. In both the field and in the laboratory where snails are left undisturbed, very few bivalves have been found with incomplete boreholes. Those which have may be explained by mechanical disturbance which dislodged the snail from its prey.

The questions thus arise whether (a) the live oyster emits some signal through the shell which continuously identifies the oyster as the predator excavates, (b) the chemical attractant from the mantle cavity provides the necessary stimulus for uninterrupted boring, or (c) contact of the snails with the shell material is sufficient for it to continue penetration without additional stimulation from the live oyster. To explore this interesting problem we carried out a series of preliminary experiments and concluded with the following one on September 10–20 (Table IV). Temperature of the seawater ranged from 20 to 18°C, and the salinity was 31 ‰.

At the beginning of day 1, 24 cleaned numbered oysters, 7–11 cm long, were placed in a grid of three rows in the experimental seawater tray used in experiments (a) to (k), and 120 *U. c. follyensis* specimens, 23–40 mm in height, were scattered among them. On day 2 we removed all snails from the tray which were not on oysters and those snails which were attached to the cupped valves of oysters. This left 21 oysters with 38 snails attached to flat valves.

On Day 3 we devalved the oysters. As gently as possible, we forced open the valves at the hinge with a stout oyster knife, severed the adductor muscle, carefully scraped off muscle adhering to the valve supporting a snail (s), and discarded the flesh and the opposing valve. The devalving operation was carried out under seawater, All macroscopic traces of adductor muscle were removed from the valve, as snails may abandon hole boring and creep around the valve to feed on the muscle. After devalving, there remained 21 oysters with 29 boring snails. Five of the snails, irritated by the mechanical grating of the oyster knife, dropped off when we forced the valves apart but resumed boring when returned to the incomplete boreholes. The remainder of the snails had excavated only shallow holes by this time and did not return. In the preliminary experiments we discovered that the rate at which snails abandoned their boreholes as a result of devalving decreased with increase of depth of the boreholes. Because a few boreholes in thinner shell were completed in 3 days, we thus chose to devalve on day 3.

Each half-shell boring-snail(s) preparation (single valve model) was then placed, inner surface up with the snail(s) suspended underneath, on a support cut from a plastic dish (Fig. 6) in the seawater tray. A binocular microscope was positioned over the tray to facilitate viewing of the final stages of penetration of the shell. Five liters of seawater flowed through the tray per minute. During daytime, standard overhead laboratory fluorescent

Table IV. Persistence in Boring by *U. c. follyensis* after the Opposing Valve and Flesh of *C. virginica* Are Removed

Oyster No.	Snail No.	Day boreholes abandoned (a) or completed (c)								Snails boring after devalving
		1	2	3[a]	4	5	6	7	10	
1	1	—	—	c						
	2	—	—	c						
	3	—	—	—	—	—	—	—	c	1
2	4	—	—	—	—	—	—	—	c	1
3	5	—	—	—	—	—	c			1
4	6	—	—	—	c					1
	7	—	—	—	—	—	—	—	c	1
5	8	—	—	—	—	c				1
6	9	—	—	—	—	c				1
	10	—	—	—	—	—	c			1
7	11	—	—	—	—	—	a			1
8	12	—	—	—	—	c				1
9	13	—	—	a						
10	14	—	—	c						
	15	—	—	—	c					1
	16	—	—	—	c					1
11	17	—	—	—	—	—	c			1
12	18	—	—	a						
13	19	—	—	a						
14	20	—	—	—	c					1
15	21	—	—	—	—	c				1
	22	—	—	—	—	c				1
	23	—	—	—	—	c				1
16	24	—	—	—	c					1
	25	—	—	—	—	c				1
17	26	—	—	—	c					1
	27	—	—	—	a					1
	28	—	—	—	a					1
18	29	—	—	—	—	c				1
	30	—	—	—	—	—	c			1
19	31	—	—	c						
	32	—	—	—	c					1
20	33	—	—	—	—	a				1
	34	—	—	—	—	a				1
	35	—	—	—	—	a				1
21	36	—	—	—	a					1
										29
Incomplete boreholes per day					3	3	1	0	0	7
Complete boreholes per day					7	8	4	0	3	22

[a] Oysters were devalved on day 3.

Fig. 6. Valve model underwater. Emerging borehole excavated by snail on underside of shell is visible to upper right of adductor muscle scar. Length of valve 6 cm. Light photograph.

lights illuminated the tray, and these were extinguished at night. Additional illumination from the microscope lamp onto the inner surface of the shell did not disturb boring snails.

The experiment demonstrated conclusively that *U. c. follyensis* can penetrate the shell of its prey in the absence of the live animal provided boring is initiated on the live prey. Thus boreholes can be completed without stimulation of any kind from the live oyster. After day 3, when devalving was performed, 22 out of 29 boreholes (76%) were completed, and seven (24%) snails abandoned their incomplete holes (Table IV). Since it was not possible to determine whether all snails in the experiment had been deprived of food equally, the less hungry ones may have abandoned their excavations because of the absence of possible reinforcing stimuli from attractant and valvular movements.

Variation in the time of completion of boreholes (Table IV) is an expression of the time when excavation was initiated, the thickness of the shell at the site of penetration, and possibly the rate of excavation of different individual snails. Snails 3, 4, and 7, for example, which did not complete their boreholes until day 10, bored through shell which was 3.4, 2.2, and 2.0 mm thick.

Under the conditions of this experiment the number of snails which completed boreholes was related in large part to the amount of mechanical disturbance inflicted during devalving. If oysters could be opened without the mechanical irritation, completion of boreholes should approach 100 percent.

2. Boring Behavior with the Oyster Model

During penetration *U. cinerea* maintains its foot firmly pressed against the shell of its prey and holds its own shell close to that of the prey, so that it is impossible from the outside to observe the process of excavation. To circumvent this visual barrier we studied boring behavior in a partially transparent artificial oyster model under binocular microscopic magnification supplemented by still and motion-picture photography (Carriker and Martin, 1965).

The oyster model consists of live shell-less oysters contained within a chamber of glass and shell (Fig. 7). *U. cinerea* is attracted to and penetrates the shell at the juncture of shell glass, where the boring operation is visible through the glass from the outside. Oyster metabolites and blood diffusing outward at the juncture provide the stimuli which attract snails to bore there;

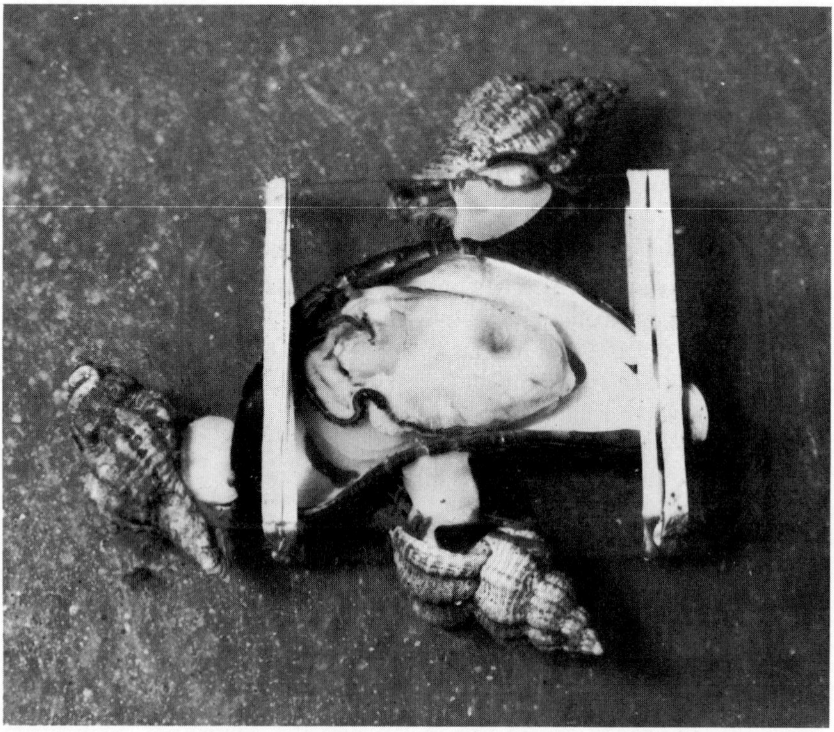

Fig. 7. Oyster model underwater. Three *U. c. follyensis* specimens are actively exploring artificial incomplete boreholes in valve of *Mytilus edulis*. Snail at left has extended proboscis to glass–shell juncture. Height of snails 38 mm. Light photograph.

empty models do not attract snails. The original concept and use of this type of viewer are credited to Prytherch (unpublished) and provided the basis for early observations (Carriker, 1943). A number of modifications have since been introduced which increase the effectiveness of the model in behavioral studies. These changes are included in the following description of the model (Fig. 7).

For snails ranging in height from 35 to 45 mm, we employed a model consisting of two glass plates 1 mm in thickness, 5 cm wide, and 10 cm long. The shell component was either the deeply cupped left valve of an oyster (*C. virginica*) or one of the valves of the blue mussel (*Mytilus edulis*) about 7 cm long and 4 cm maximum width. Because of the uniformity and depth of mussel valves and the contrast afforded the organs of the snail by the blue-black color of the shell, we utilized blue mussel valves most often. The free margin and opposing middle portion of the valve were ground flat on wet silicon carbide paper of grit ♯240 on a rotary Buehler wheel to provide a shallow collar about 1 cm in depth. The coarse grit impressed a pattern of scratches on the shell surface at the point of contact with the glass plate sufficiently coarse to allow diffusion of metabolites across the juncture detectable by the snails.

Repeated observations by us that snails are able to detect minute chinks broken accidentally at valve junctures or elsewhwere over the otherwise intact shell of live bivalves suggested a means of increasing the predictability of attack by snails at given sites on the model. Accordingly, we drilled several incomplete boreholes, about 1 mm in diameter and 1–2 mm deep, with a small electric drill in the shell at the shell–glass juncture (Fig. 7). We found that snails locate these holes and excavate the shell at these points in preference to nondrilled points along the shell–glass juncture, probably attracted by metabolites moving out of the model more rapidly there then elsewhere.

The two glass plates were secured to the shell collar by means of two stout rubber bands. Elasticity of the bands reduced breakage of the plates and permitted gentle intermittent manual elevation of one of them from the shell collar (opposite the boring snail) to promote exchange of seawater inside the model and thereby extend the life of the oysters. The short space between the glass plates allows the snail to extend its head and foot, but not ordinarily the shell, to the shell–glass juncture, thus permitting a clear view of the operation of the boring organs (Fig. 8).

The interior of the model was filled with one to four fresh, live, actively metabolizing oysters 25–45 mm in shell length, removed from their valves by careful excision of the adductor muscle at the shell attachment sites. Oysters were injured as little as possible to minimize bleeding. Assembly of the model was carried out under seawater to eliminate trapping of air bubbles. The model was immersed in a glass bowl approximately 8 cm deep

Fig. 8. Limited space between glass plates of oyster model allows *U. c. follyensis* to extend soft parts but ordinarily not the shell, permitting a view of operation of boring organs. Proboscis 1.5 mm in diameter is supported between propodial ridges. Light photograph.

and 20 cm in diameter, and six to eight snails were positioned along the long side of the model with siphon tips pointing between the glass plates toward the incomplete artificial boreholes. The glass bowl was then submerged in a shallow tray of running seawater to a depth allowing a layer of water about 1 mm in thickness to flow gently over the top edge of the bowl. The flowing seawater provided a gentle exchange with that in the bowl, kept oysters alive within the model for 2–3 days at a time at about 20°C, and provided conditions in which snails were attracted to the artificial holes and excavated them further.

After boring commenced we transferred some bowls from the running seawater tray onto a bench under a cantilevered binocular microscope for detailed observations and for photography. Other models were observed *in situ* in the tray with the microscope and camera mounted over the water. After 1–3 days, because of deterioration of the oysters and reduction in rate of release of metabolites, attractiveness of the model decreased rapidly. It was thus not possible to observe sequences of penetration for more than 1–3 days at a time, longer times being favored by lower temperatures. Occasionally, deep artificial boreholes were completed by snails. Rate of penetration of shell in the model was slower than that in normal prey shell, probably because of diffusion of seawater into the borehole at the shell–glass interface and because of imposition of the glass plate over the hole. In spite of these impediments, however, the model proved invaluable in the study of several aspects of the shell-penetrating behavior of *U. cinerea*.

Ambient seawater containing what appeared to be higher concentrations of attractant than that diffusing from the oyster model sometimes masked the attractiveness of the model. This difficulty was particularly troublesome at the Institute of Fisheries Research, where the seawater intake fouled rapidly. By hand-hauling seawater from the bay and artificially aerating it for an hour or so, or filtering it through clean fine sand, we were able to reduce its attractiveness so that snails would respond to the model (Blake, 1962). This problem was not encountered to a detrimental degree at the Marine Biological Laboratory because of less fouling of seawater pipes and the greater flow of seawater.

In an attempt to improve the attractiveness of the oyster model, we introduced sand-filtered seawater into an oyster-filled model by means of a minute tube in the shell collar and let it escape slowly at the shell–glass interface. This stimulated snails to explore the glass top of the model, a predictable response in view of the habit of snails to avoid the edge of the valves of pumping oysters and to bore away from the valve margins. When flow of water through the model was discontinued, the snails again responded to the attractant at the juncture. Further attempts to improve the model included (a) packing the interior of the model with small live whole oysters and (b) slowly piping water from actively filtering oysters in a reservoir into an empty model. Innovation (a) provided longer-lived "bait" which would not decay, but the confined conditions within the model soon caused the oyster spat to close. Although snails aggregated at the shell–glass juncture receiving effluent from pumping oysters (b), they did not attempt to bore.

For purposes of still and motion photography of the boring process, we set up four to six oyster models in the early morning and by late afternoon or early evening a few snails were actively boring. After snails had been boring for several hours, occasional use of photographic lights did not appear to inhibit or alter the boring behavior.

3. Boring Behavior

Shell penetration involves the close interaction of the proboscis, propodium, and accessory boring organ (ABO), in this order, in a predictable cycle which repeats itself continuously throughout the process of boring of each borehole. The duration of each phase of each cycle is relatively constant except at the beginning and completion (breakthrough) of each borehole.

U. cinerea and the subspecies *U. c. follyensis* approach the shell–glass juncture of the oyster model in various positions, depending upon the portion of the model first contacted by the propodium. As the foot creeps between the glass plates onto the shell collar, the propodium is passed slowly over the shell–glass juncture, minute waves rippling across the anterior ventral margin. This peristaltic-like activity is characteristic of the searching behavior of this part of the foot during shell penetration and is suggestive of a tactile function. The fact that the inhalant siphon of the snail often is forced to remain outside the glass plates and the propodium is frequently the first to move to the juncture supports in addition a chemosensory function for it (Fig. 8). In the early stages of exploration the snail frequently extends its proboscis openly to the juncture. Passing it back and forth over the juncture and neighboring shell and glass surfaces, and as it does so characteristic peristaltic-like waves visible only under the microscope move rapidly back and forth over the peristomal rim. From time to time during exploration, the mouth opens and the buccal cavity enlarges in what appears to be a "tasting" reaction. In a modified oyster model (Carriker *et al.*, 1967) we injected minute quantities of a vital dye over the incomplete borehole, and it was possible to see the snail take seawater into the buccal cavity during this "tasting" activity. In the approach phase the snail rasps occasionally, and the proboscis is sometimes extended its full length, assuming a variety of curious postures imposed by the model. If the rubber bands securing the model are not sufficiently strong, the snail will force the proboscis through the juncture, sections of the proboscis pulling themselves worm-like through the narrow slit. The portions before and after the constriction ballooning from blood pressure (Fig. 9). Anterior propodial ridges are used only partially,

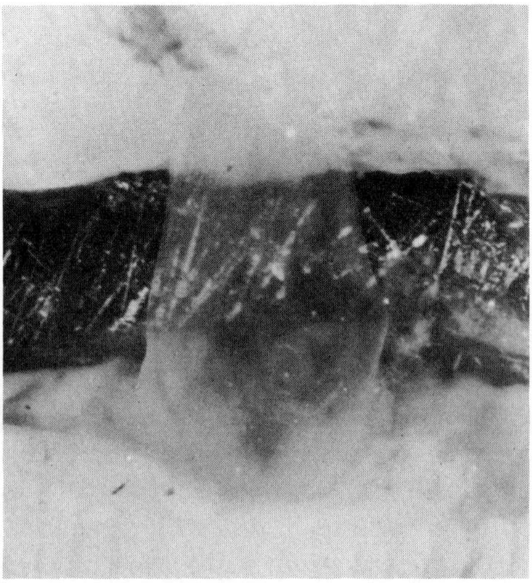

Fig. 9. Proboscis of *U. c. follyensis* (snail at top) squeezed between glass and shell of oyster model, feeding on oyster gill (bottom). Coiled structure below shell section is radular sac. Diameter of unflattened proboscis 1.5 mm. Light photograph.

and sometimes not at all, in supporting the proboscis during search for a penetration site.

After the penetration point has been chosen at the shell–glass juncture or in an incomplete artificial borehole, the snail rasps and swallows the loose material at the shell surface, secures a strong hold on the substratum with its foot, and begins penetration. The position of the snail shell relative to that of the model does not generally change during perforation, thus discrediting the popular notion that boring is accomplished by rotation of the shell about the borehole. The snail does, however, periodically slightly raise, lower, and move its shell sideways during excavation.

Actual shell excavation is accomplished by an intricate chemical–mechanical process in which the ABO secretes a substance which dissolves shell at the site of penetration, and a minor portion of the weakened shell is then removed by the radula and swallowed. Alternation of successive short periods of rasping and long periods of chemical activity continues until a borehole is excavated.

a. Rasping of Shell. The mouthparts of *U. cinerea* are contained within the distal end of the long muscular tube-shaped proboscis which is an extension of the head region. While inactive the proboscis is housed, base first, within the head. Fully extended by combined muscular activity and blood pressure, the proboscis may slightly exceed the height of the shell (Carriker, 1943).

The radula is contained within the buccal mass at the distal end of the proboscis and is supported on the odontophore within the buccal cavity (Fig. 10). From a long tubular radular sac back of the odontophore the radula opens out of the sulcus, spreads forward over the front, the bending plane, and then extends backward a short distance under the odontophore. The radula consists of three longitudinal rows of nearly colorless mineralized teeth supported on a radular membrane (Fig. 11). In the resting position the teeth overlap and point posteriorly. Each transverse row of teeth consists of a sturdy central rachidian tooth and two slender marginal teeth. Each rachidian tooth possesses five prominent cusps arising from a stout basal plate attached to the radular membrane. Unworn cusps are sharp, slightly hooked, and curve posteriorly. Each rachidian tooth is hinged at its posterior basal edge to the radular membrane. This permits the teeth to overlap posteriorly when at rest. At the bending plane, each tooth is rocked forward until its flattened base rests against the taut radular membrane, supported at an angle at which the cusps are upright in the effective rasping position. Each unicusped scythe-shaped marginal tooth is attached primarily on the anterior side of the tooth. Upon erection at the bending plane, the distal half of each tooth points upward from the membrane. As the radular membrane overfolds into the sulcus, the marginal teeth incline inward over the rach-

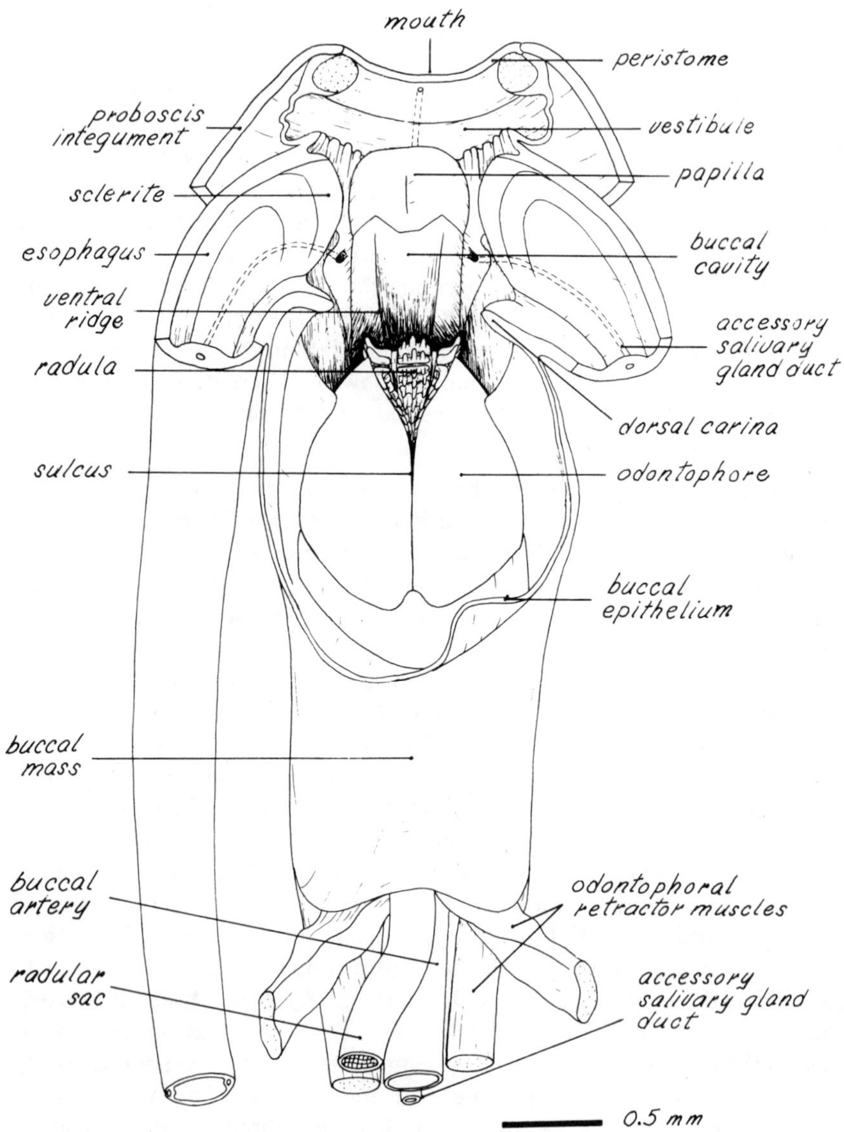

Fig. 10. Drawing of dorsal view of buccal mass and esophagus of *U. c. follyensis* opened along median sagittal line to expose internal surfaces and openings of salivary gland ducts and accessory salivary gland duct.

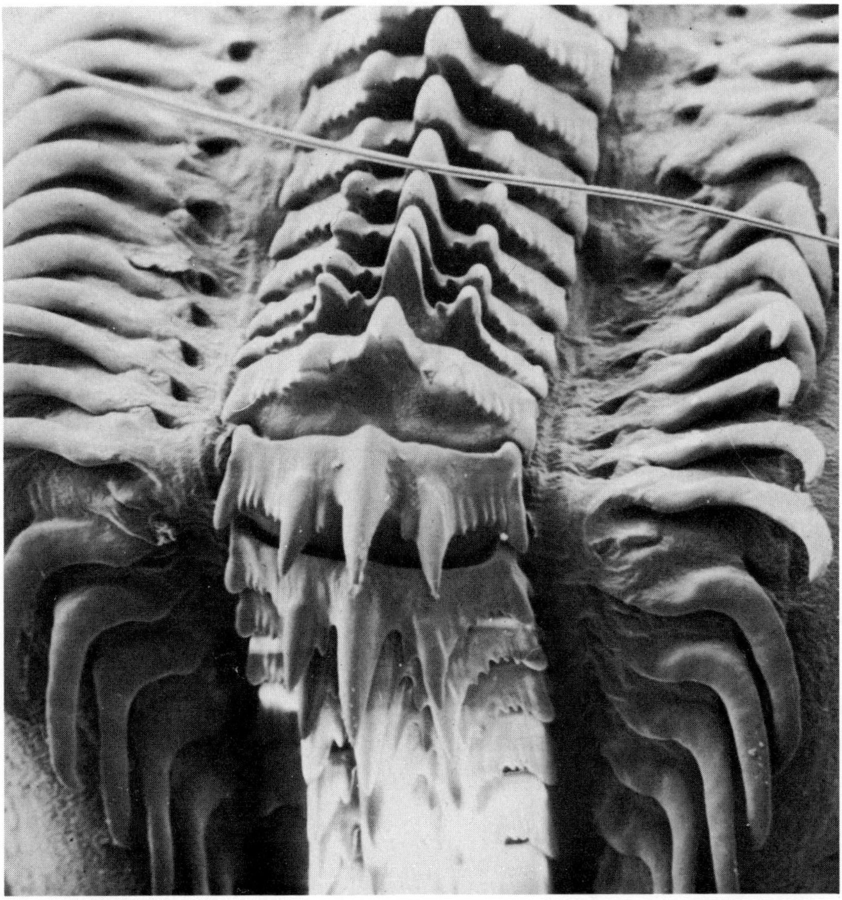

Fig. 11. Radula of *U. c. follyensis* at bending plane, showing central row of rachidian teeth and two rows of marginal teeth. Width of the radula 410 μ; snail height 40 mm. Scanning electron micrograph.

idian teeth. The width of a transverse row of teeth on the unfurled radula at the bending plane in a snail 40 mm high may be as much as 400 μ, but radular dimensions vary considerably among different individuals of the same shell height (Carriker, 1969).

The radula is slid back and forth over the front ends of the twin odontophoral cartilages by the musculature of the subradular membrane, which is held tautly by blood pressure within the odontophoral sinus. Teeth are erected as the radula slides anteriorly out of the sulcus. Motion pictures have demonstrated that during the shell-rasping stroke the anterior half of the odontophore, closely surrounded but not generally touched by the peristome, is extended beyond the mouth, sharply ventrally toward the substratum. Upon contact of the cusps with the shell surface, the anterior end of the odontophore is moved forward in a "licking motion," scraping the cusp points across the surface, and then the odontophore is taken back into the buccal cavity. As the radula is drawn over the odontophore and folds into the sulcus, shell flakes removed by the rachidian teeth are drawn by suction into the esophagus. During passage of the odontophore out of the mouth in readiness for the next abrasive stroke, the radula is simultaneously pulled forward over the cartilages. As the radula makes contact with the substratum, its movement is reversed. Thus during the effective rasping stroke, both the odontophoral tip and the radula move over the shell surface, but the radula at a faster rate than the odontophore (Fig. 12). This distributes abrasion of the radular cusps over several transverse rows of teeth and increases the efficiency of rasping. Cusps blunt progressively toward the anterior end of the radula. Worn terminal teeth are dislodged during rasping and swallowed. The radula, however, grows forward gradually out of the radular sac, so that new teeth continuously assume the rasping function at the bending plane (Carriker, 1969; Runham, 1963).

When first protracted from the mouth, the anterior portion of the odontophore takes the shape of a tapering spoon-shaped scoop, with the single longitudinal row of rachidian teeth positioned over the middle of the crest of the bending plane. As the radula rasps over the shell surface, the front end of the odontophore broadens noticeably. Rachidian teeth do not appear to bend, so that intervals between the points of the five cusps remain relatively constant as the teeth ride over the bending plane. Because of the

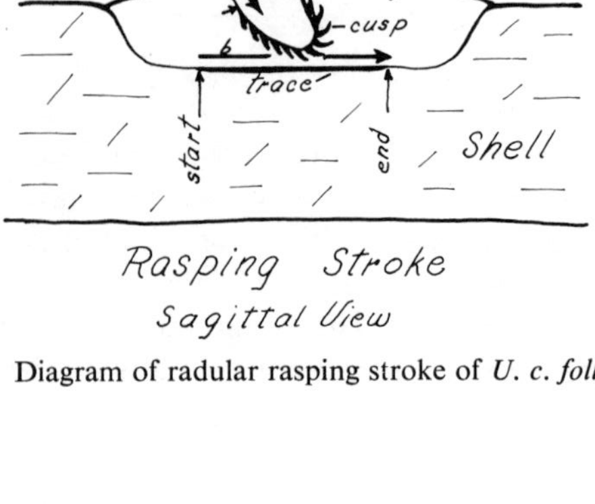

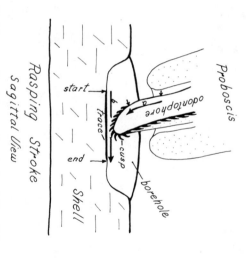

Fig. 12. Diagram of radular rasping stroke of *U. c. follyensis*.

unfolding of the radula beyond 180°C at the bending plane, however, only rachidian teeth generally come in contact with the shell surface with any force. Consequently, marginal teeth exhibit relatively little wear. The primary function of the marginal teeth, working synchronously with the rachidian teeth, is to tear off bits of flesh when feeding (Carriker, 1969) (Fig. 17).

The radula is ineffective in rasping normal hard shell of bivalves. A critical question relative to penetration of shell by boring gastropods has been whether radular cusps are harder than the shell of their prey. Tests made with a microhardness tester demonstrated that marginal teeth of *U. c. follyensis* were about twice as hard (average approximately 400 kg/mm^2) as rachidian teeth (200 kg/mm^2). The range of hardness of rachidian teeth was similar to that of the shell of *C. virginica* (Carriker, 1969). Although the relative softness of the rachidian cusps was surprising in view of their function in hole boring, it is logical in terms of their function in tearing flesh that the slender, more easily broken marginal teeth should be the harder of the two.

We next sought an explanation for the differential in hardness of marginal and rachidian cusps by examination of the elemental composition of the teeth. We excised the radulae of seven *U. c. follyensis*, leaving some of the muscle adhering to the radular shield. In a drop of distilled water, after rinsing each radula several times, we folded the forward third under, leaving a transverse row of relatively unworn marginal and rachidian teeth projecting freely. Each radula was mounted on polyethylene with Duco cement, the adhesive being applied some distance back from the projecting teeth to avoid contamination. Snails ranged in shell height from 37 to 40 mm, the radular width from 235 to 333 μ, and the width of rachidian teeth from 118 to 137 μ. Mr. James Bussey of Jarrell Ash Division, with our help on orientation of cusps, then made analyses with the laser microprobe of two to three cusps on each projecting transverse row of teeth. Each sample was vaporized to a useful spectral emission by the laser beam, which was concentrated onto the sample by a microscope objective lens, and the elements present in each sample were determined by comparing the spectrum of the unknown with that of standards of pure samples. A total of eight rachidian and seven marginal cusps was analyzed. Rachidian cusps included one or two of the three central cusps on each tooth. Preliminary tests had been carried out on a variety of possible mounting media in a search for materials with minimum background contamination. The laser beam burned a pit about 15 μ in diameter and as deep (the approximate dimensions of the base of each of the three central major cusps of rachidian teeth, and slightly larger than the basal portion of marginal teeth). The beam may thus have been slightly larger than some of the samples. As calcium was the major constituent in the teeth, we adjusted the value of all the other elements to an average calcium percentage

Table V. Elemental Analysis of Radular Cusps of *U. c. follyensis* by Laser Microprobe[a]

		Element								
		Ca	Sr	B	Mg	Si	Zn	Fe	P	Ti
Rachidian cusps (8)	Range	H–H	T^+–H	T^+–M^+	T^+–M	T^+–H	T^-–M	T^-–T^+	N–T^+	N–T^+
	Average	H	M	T^+	T^+	T^+	T^+	T^+	T	T^-
Marginal cusps (7)	Range	H–H	T–M^+	T^+–M^+	T^+–M	T^+–H^+	T^-–M	T^-–M	T^-–T^+	N–T^+
	Average	H	M	M	T^+	T^+	T^+	T^+	T	T^-

[a] H = major constituent, M = minor constituent, T = trace constituent, N = not detectable.

transmission of 30 (range in different cusps 16–48). The tests were qualitative, but a rough index of concentration was somewhat as follows: major constituent, over 5–10%; minor constituent, 0.1–5%; trace, less than 0.1%.

The results of analysis of elements in the radular cusps are recorded in Table V. There was a wide range of variation in the proportion of elements in cusps of the same radula and in different radulae, but the average elemental composition of both rachidian and marginal teeth was similar. Strontium and silicon ranged from major to trace; boron, magnesium, zinc, and iron from minor to trace; and phosphorus and titanium from not detectable to trace. An explanation for the structural difference in hardness between the rachidian and marginal teeth of *U. c. follyensis* must thus be sought in other characteristics of the teeth, possibly in the crystal structure.

These data confirm an earlier observation which demonstrated that silicon is not present in sufficient amounts in the radula of this species to preserve the form of the teeth in boiling concentrated sulfuric acid (Carriker, 1943). Runham (1961) also reported iron and silicon in the teeth of the nonboring gastropod *Patella vulgata.*

b. Chemical Dissolution of Shell. The chemical component of the boring mechanism of *U. c. follyensis* is provided by the ABO. In the inactive state, this organ is contained within a combined vestibule and large sinus in the median anteroventral region of the foot (Fig. 13). When active, the ABO is pressed firmly into the borehole, its diameter and length approximating those of the borehole. Within the excavation, as seen in the oyster model (Fig. 18), the ABO takes the form of a thin-walled translucent cylinder with a slightly convex apical secretory disc covered by a brush border. The disc ranges in color in different snails from cream to orange-brown. Withdrawal of the ABO from the hole is accomplished by contraction of muscles passing from the foot through the sinus to the proximal surface of the disc. Arborescing arterioles, nerves, and muscles originating in the cephalic region and in the foot penetrate the epithelium freely through the ABO sinus. Secretory cells occur as groups surrounded by small sinuses within the disc. Muscle strands interweave among the cells and abut the free end of the secretory epithelium. Blood flows into the epithelial sinuses by way of minute capillaries and thence into the major ABO sinus. Nerves branch profusely throughout the epithelium and form a complex nerve plexus among the secretory cells. These cells contain dense populations of mitochondria and a variety of granules. It is not yet known which of the granules is the precursor of the shell-dissolving substance (Carriker, 1969; Nylen *et al.*, 1969).

Dissolution of shell occurs slowly on contact of the ventral surface of the secretory disc of the live intact ABO with shell (Carriker, 1961; Carriker *et al.*, 1963). Secretion from the disc is a viscid acid substance with a pH of about 3.8–4.1 (Carriker *et al.*, 1967). Cytochrome oxidase, succinate dehydro-

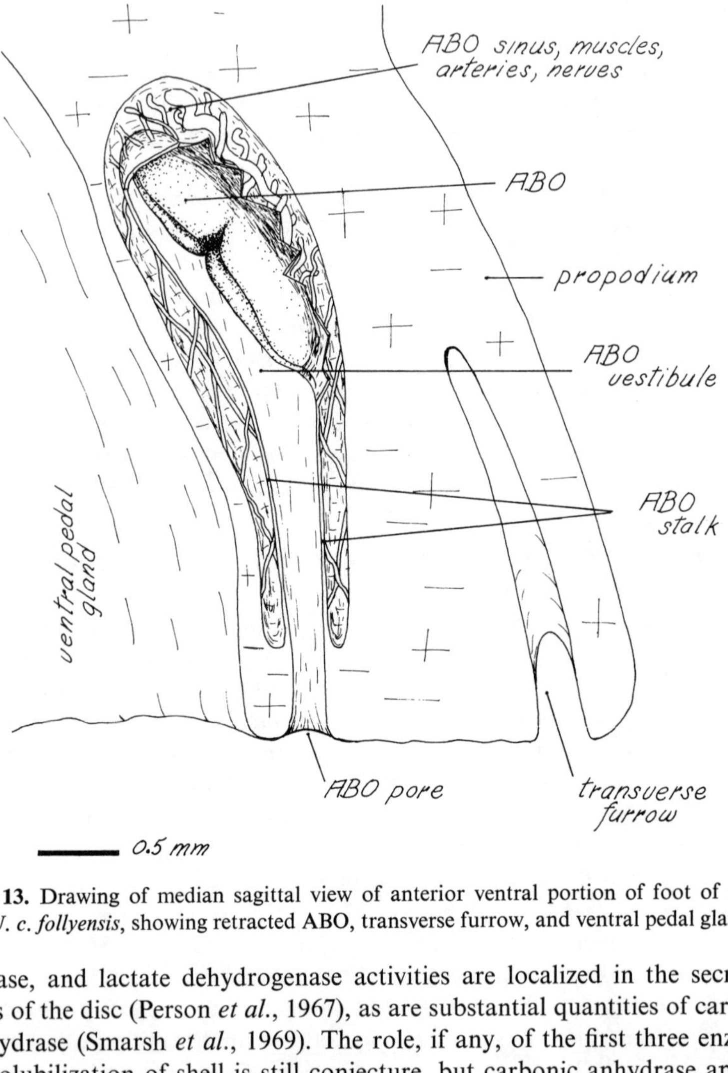

Fig. 13. Drawing of median sagittal view of anterior ventral portion of foot of female *U. c. follyensis*, showing retracted ABO, transverse furrow, and ventral pedal gland.

genase, and lactate dehydrogenase activities are localized in the secretory cells of the disc (Person *et al.*, 1967), as are substantial quantities of carbonic anhydrase (Smarsh *et al.*, 1969). The role, if any, of the first three enzymes in solubilization of shell is still conjecture, but carbonic anhydrase appears to be an important link in the mechanism. The nature of the chemical mechanism is still unknown.

The anatomy and histology of the ABO and the foot of the snail suggest that the shell-dissolving secretion arises entirely within the secretory disc of the ABO. To check the matter, we tested the shell-dissolving capacity of excised ABOs and other representative tissues in the snail body. The snail, after removal from its shell by cracking, was cut in half at a frontal plane between the tentacles and the foot. The foot was then pinned upside down onto a rubber eraser embedded in a dissecting pan, the dorsal region above the ABO positioned over a smooth metal sphere about 1.7 mm in diameter secured to the surface of the eraser. Pressure from iris scissors and fine forceps pressed the ABO out of the foot, and the stalk was cut free close to the disc. The excised ABO was then placed, secretory disc down, in a small drop of seawater, on the surface of a square of shell (*Spisula solidissima*) ground and polished to a nearly scratchless surface (Carriker and Van Zandt, 1964) and covered with a disc of thin transparent plastic sheeting ("Handi-wrap") which by capillarity applied a gentle pressure to the back of the gland (Fig. 14). The ABO–shell preparation was housed in a moist chamber at room temperature. After periods ranging from 15 min to 32 hr, the ABO was rinsed off and the shell surface was air-dried, coated with chromium in vacuum, and examined microscopically by polarized incident illumination.

The technique demonstrated that secretion from the majority of excised ABOs etched the shell, often leaving conspicuous imprints which faithfully mirrored the outline, shape, and creases of the disc (Figure 15). Control tissues from other parts of the body did not etch. The active agent was not removed by rinsing in seawater and continued to be secreted by the ABO up to an hour of soaking in seawater. Etching also took place in a CO_2-free atmosphere. The majority of halves of active ABOs cut apart with a sliver of a razor blade over teflon each etched polished shell as conspicuously as intact organs (Fig. 16). The ABO-shell preparation provides a useful qualitative method for determining the etching capacity of whole or portions of ABOs and supports the hypothesis that the secretion arises within the secretory

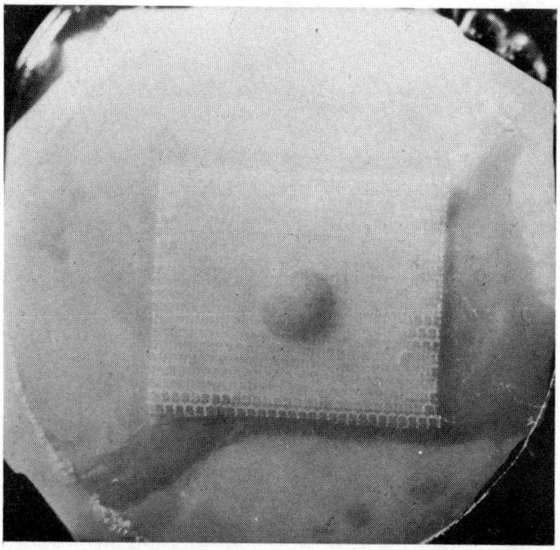

Fig. 14. Excised accessory boring organ of *U. c. follyensis* placed disc down on plankton silk over polished shell (*Spisula solidissima*) under plastic sheet to test shell-dissolving capacity of gland. Diameter of gland spread by pressure of sheet to 3 mm. Light photograph.

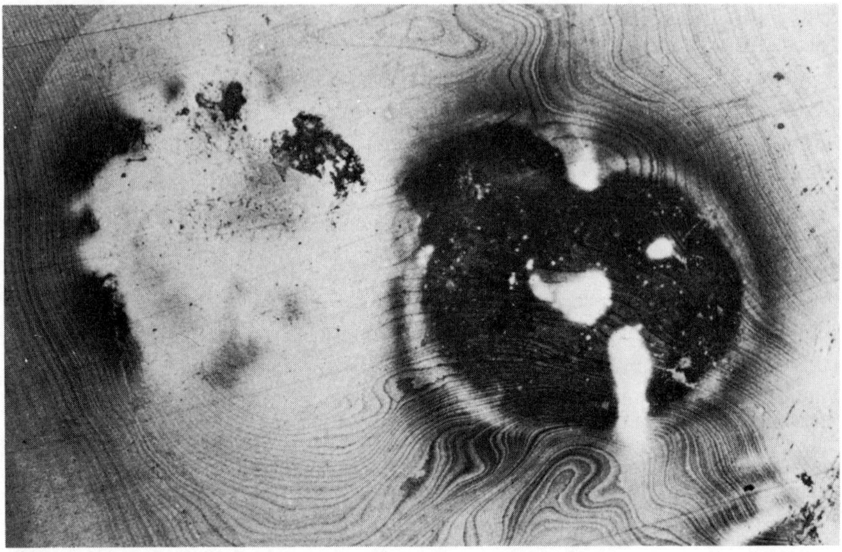

Fig. 15. Two dissolution patterns of excised accessory boring organs of *U. c. follyensis*. ABO to left was inactive, etching only small region along left circumference of gland. ABO on right was active and except for four spots (white) etched shell deeply. The pattern of "contour lines" represents growth rings in shell exposed by secretion. Under inactive portions of glands no etching occurred. Diameter of each pattern about 3 mm. Light photograph.

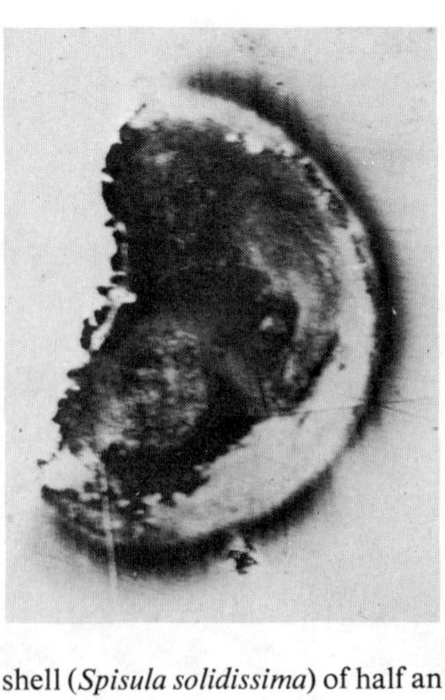

Fig. 16. Imprint of polished shell (*Spisula solidissima*) of half an excised ABO. Imprint 3 mm long. Clean edge to the left was the cut edge of the gland. Light photograph.

disc. Why some organs out of every sample of snails are inactive is unexplained and does not appear to be related to the resting or boring state of the snails.

c. Interaction of Proboscis, Propodium, ABO, and Shell. Prior to the beginning of penetration, the snail positions itself on the shell of its prey with the pore of the retracted ABO located over the prospective boring site. The pedal epithelium thereafter generally remains in the same position and firmly attached to the shell surface. This clears up an early conjecture that the snail moves back and forth as the proboscis and ABO are alternated (Carriker, 1943, 1955*a*). The stationary position obviates the need for the snail to relocate the borehole by movement of the entire foot after each rasping period. The anterior medial portion of the propodium is then retracted deeply and the lateral propodial ridges are overfolded, forming a fleshy tube over the borehole site down which the proboscis is extended (Fig. 8). The pedal tube admits seawater as it forms which bathes the borehole site and lubricates rasping. Rasping is limited principally to the bottom of the incomplete borehole (Fig. 17, bottom). The odontophore is rotatable on its lengthwise axis independently of rotation of the proboscis by at least 180°. Thus by swinging to the left and then to the right in two half turns, the odontophore can cover the circumference of the borehole; the additional rotational capacity of the proboscis contributes further flexibility in rasping. The radula is pressed against the substratum with considerable force, noticeable through the glass of the oyster model (Fig. 17, middle). Before and after the rasping period, the snail generally passes the peristome close over the surface of the borehole in slow exploratory movements for a few seconds to about a minute. The peristome is liberally supplied with nerves and probably is important in chemosensory and tactile activities.

At the close of the rasping period, the proboscis is withdrawn into a sac in the cephalic hemocoel, and as this is done additional seawater flows into the incomplete borehole. Simultaneously, the midanterior portion of the propodium, already at the posterior edge (relative to the orientation of the snail) of the borehole where it surrounded the proboscis, is extended into the borehole and, pressing the partially flared transverse furrow (Fretter and Graham, 1962) closely against the surface of the borehole, is slid forward across the bottom and up the anterior wall of the borehole (Fig. 18, top) and back onto the surface of the shell to assume a normally extended position as well as a tight contact between the epithelium of the snail's foot and the prey's shell. In this maneuver the propodium voids seawater from the incomplete borehole prior to the entrance of the ABO, thus minimizing dilution of the secretion of the ABO. This is important in view of the chemical nature of this phase of boring. The role of the transverse furrow is more difficult to explain. The presence of the large anterior pedal mucous gland in the

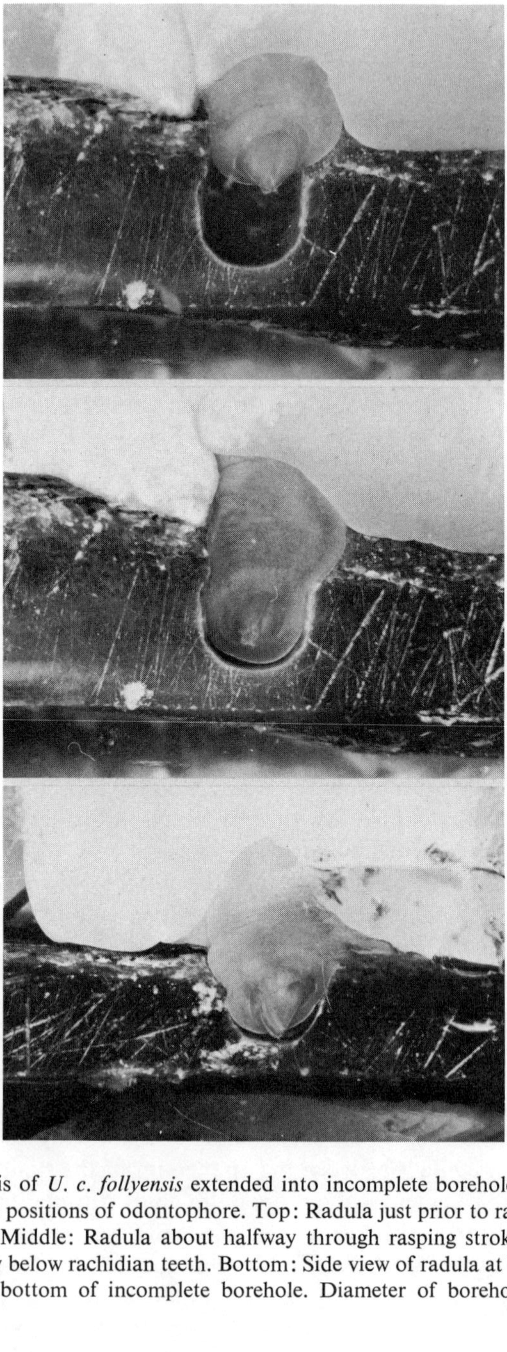

Fig. 17. Proboscis of *U. c. follyensis* extended into incomplete borehole in oyster model, showing different positions of odontophore. Top: Radula just prior to rasping across glass plate over hole. Middle: Radula about halfway through rasping stroke, marginal teeth depressed slightly below rachidian teeth. Bottom: Side view of radula at beginning of rasping stroke over bottom of incomplete borehole. Diameter of borehole 1.5 mm. Light photographs.

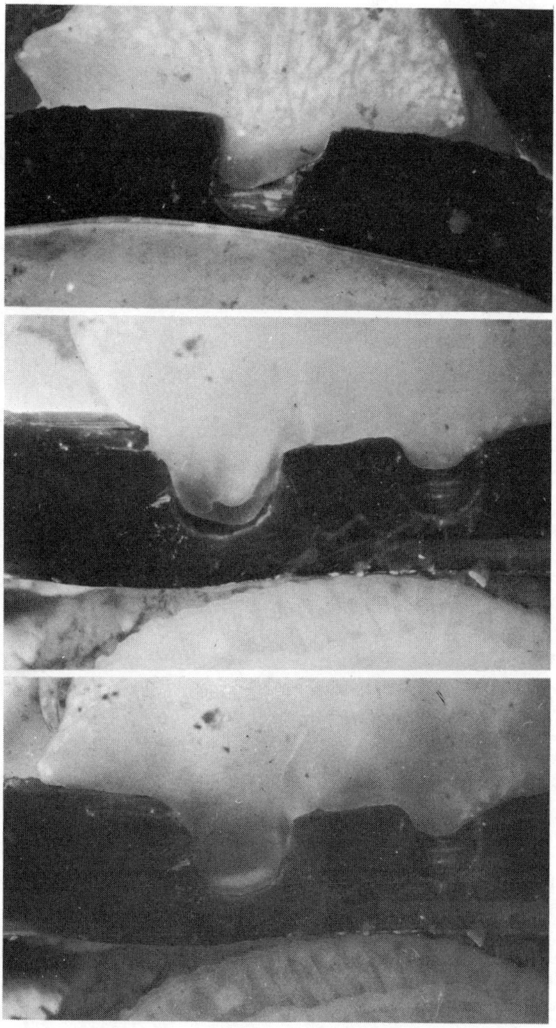

Fig. 18. *U. c. follyensis* on incomplete borehole in oyster model. Top: Front view of snail, propodium passing up anterior wall of the borehole (against glass of model), pedal region around ABO beginning to extend into hole. Middle left hole: Folded stalk of the ABO moving into borehole in advance of the ABO. Middle right hole: Bulge of foot reaching into borehole. Bottom: ABO in position in borehole. Middle and bottom photographs are side views of snail. Diameter of larger borehole 1.5 mm. Light photographs.

anterior part of the foot, discharging through a canal into the midregion of the furrow, suggests a lubricatory function. Whether the secretion from this gland has a role in initiating dissolution of the shell prior to each application of the ABO is not known. The propodium also serves to push aside partially dissolved shell. Tiny windrows of this are sometimes visible at the sides of incomplete boreholes (especially in the soft shell of *Mya arenaria*) when boring is interrupted suddenly.

It takes the propodium about 2–3 sec to move across the bottom of the borehole, passing from posterior to anterior. The propodium is followed immediately by the ABO, which slides gently into position, pressing closely against the shell surface (Fig. 18, bottom). In the retracted position (Fig. 13), the ABO lies obliquely with reference to the surface of the foot. As the ABO is extended pliantly into the borehole, the lower (or morphologically anterior) part billows sideways into the hole first, partly clothed by a fold of the retracted stalk epithelium. Frequently, the folded stalk follows the ABO a short distance into the borehole, providing the fragile microvilli some protection from abrasion against the sides of the hole (Fig. 18, middle). In other individuals the stalk may unfold early and the ABO slide unprotected into position. Under the light of the microscope the surface of the disc glistens as the ABO accommodates to the contours of the borehole, the sheen resulting from secretion clinging to the surface of the gland. Once in position, the ABO continues to undergo gentle pulsatory movements at the rate of about 20–30 per minute at approximately 20°C. When a snail with the ABO extended into a deep borehole is pulled gently off its prey, the ABO comes away with some resistance, making a barely audible sucking sound, suggesting the snugness of fit.

During its stay in the borehole, the ABO secretes sufficient solubilizing fluid to remove a thin layer of shell at the bottom. This is generally enough to obliterate most of the rasp marks of the previous rasping periods. Using the scanning electron microscope, we examined the effect of the secretion on the shell of the oyster (Carriker, 1969). The secretion exposed the gross microscopic morphology of the shell (Fig. 19), revealing etched subsurface prisms which corresponded in approximate size and form with the distinctively shaped normal prisms of oyster shell. These observations suggest that the secretion initially penetrates shell through the nonmineralized organic matrix of the prisms.

The prolonged closeness of fit of the ABO in the borehole suggests that dissolved shell products are transported across plasma membranes into the snail in exchange for other ions. Preliminary studies by Chetail and Fournié (1970) suggest that calcium ions enter the ABO while it is active. Dissolved products remaining in the hole at the end of the ABO phase of the

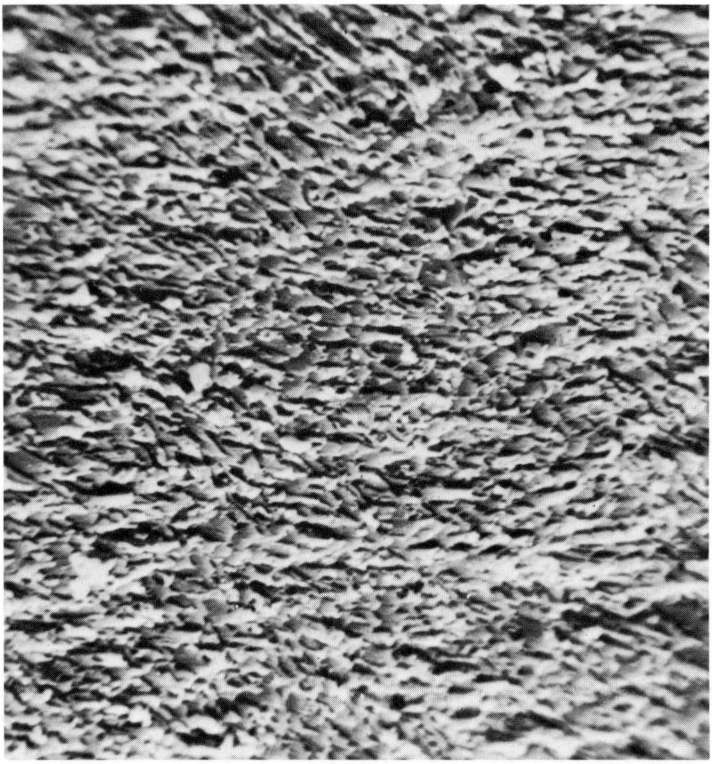

Fig. 19. Surface of bottom of incomplete borehole in shell of *C. virginica* after chemical activity by ABO of *U. c. follyensis* for about 20 min. Overlapping, pitted, shingle-like structures (largest ones in center about 1.8 μ wide) are partly dissolved shell prisms. Scanning electron micrograph. 2900×.

cycle may also be pressed out by the propodium in its transhole excursion. The proportion of voidance by each process is not yet known.

Withdrawal of the ABO from the borehole after the period of dissolution consumes about a half second. Simultaneously, the pedal tube is formed, the proboscis is extended down it to the borehole to resume rasping, and a new penetration cycle commences. Rarely, the proboscis may be inserted into the borehole simultaneously as the ABO is withdrawn (Fig. 20). Secretion remaining on the borehole surface may not only affect the teeth but may also serve a purpose in rasping. These possibilities have not been investigated.

As excavation approaches the inner surface of the shell of a valve model, there is present before breakthrough a thin, often nearly transparent sheet of shell through which the activities of penetration are clearly visible. The initial break may be made by chemical dissolution or by rasping. The snail senses the tiny hole by both the propodium and the proboscis. As the break enlarges, the phases of each cycle of penetration become modified as follows: (a) the duration of the rasping periods and the patterns of rasping become more variable, (b) the proboscis continues to test the hole for fit until it is able to pass through it, (c) the propodium is forced deeply through the hole, and (d) the duration of the ABO bulging out of the hole becomes more variable.

The pattern of rasping is especially clearly visible under low binocular magnification just before breakthrough in the purple-colored adductor muscle of oyster shell, which contrasts with the whitish rasp marks as they are made. The length of each mark is about one fourth to one fifth the outer diameter of the borehole. The pattern of rasping is quite variable and generally asymmetric. The radula may rasp back and forth across one side of the bottom leaving more-or-less parallel, roughly equidistant marks (Fig. 21); or it may scratch at roughly right angles across the first marks; or it may follow the circumference of the hole; or it may move across the bottom of the hole, back across in the reverse direction, and then up onto the side of the wall of the hole and back down again partly across the bottom. Observations of rasping patterns made on valve models confirmed those made by interrupting snails at the end of the rasping period (determined with the contact hydrophone) while boring live oysters. Boreholes were coated with chromium in vacuum and examined with incident illumination (Carriker, 1969). In valves in which hardness and possibly other characteristics are distributed hetero-

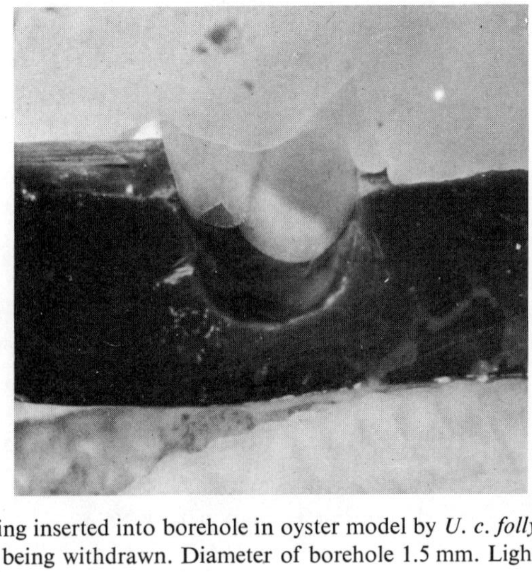

Fig. 20. ABO being inserted into borehole in oyster model by *U. c. follyensis* at same time proboscis is being withdrawn. Diameter of borehole 1.5 mm. Light photograph.

Fig. 21. Rasp pattern of *U. c. follyensis* in shallow incomplete borehole in shell of *Mya arenaria*. Relatively soft shell of *Mya* illustrates rasp marks more clearly than harder oyster shell. Long axis of borehole 1.5 mm. Light photograph.

geneously, the form of the borehole is modified by the variable structure of the shell material. This is illustrated by the longitudinal section of a borehole excavated in the valve of an oyster which was alternately stratified by soft (chalky) and hard layers (Fig. 22, bottom). The diameter of the hole was noticeably larger in the chalky than in the hard strata. In homogeneous shell the borehole was relatively uniform (Fig. 22, top). Boreholes were cut in two with a high-speed diamond wheel and coated with chromium in vacuum to provide contrast.

Once a break has been made in the bottom of the incomplete borehole by dissolution, transforming the bottom into a thin circumferential ledge, some snails rasp at the edge and the walls of the hole, ignoring the perforation. Other snails vigorously attack the edge with the radula, breaking off bits of shell and swallowing them, the force of the radula being strong enough momentarily to catch cusps on the edge and stop the movement of both the radula and the odontophore. The shape of the enlarging break at the bottom of the borehole varies widely, mainly as a result of the architecture of the shell at the point of penetration, and may range from circles, ovals, and crescents to irregular outlines (Fig. 23 and Fig. 24 top left) (Carriker and Yochelson, 1968). Rasping during completion of the perforation may also vary, ranging from a few rasps to over 100 per rasping period.

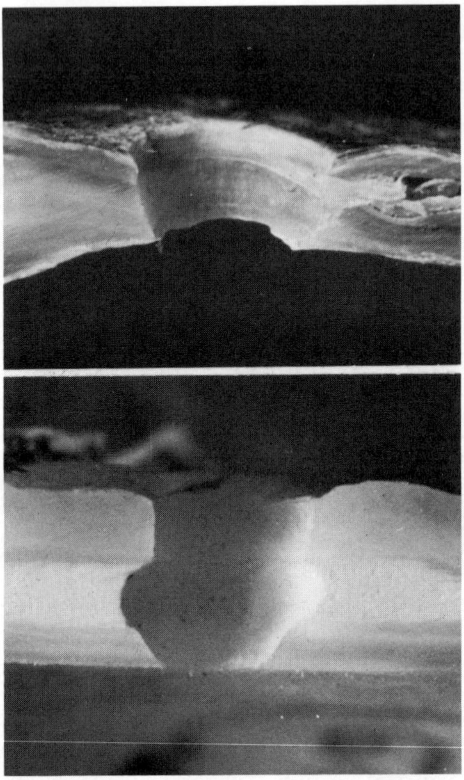

Fig. 22. Longitudinal sections of boreholes excavated by *U. c. follyensis* in shell of *C. virginica*. Top: Relatively homogeneous hard shell (borehole made by snail in recording 2). Bottom: Chalky stratum sandwiched between an outer (upper) and an inner (lower) stratum of hard shell. Outer diameter of boreholes about 1.5 mm. Light photographs.

The radula in most nonboring gastropods functions primarily in obtaining food. In shell-boring snails the secondary function of scraping shell is added. This role, however, is quite subordinate to chemical dissolution by the ABO, and the radula serves mainly in removing shell fragments and debris, particularly at the beginning and termination of excavation, and in scraping the surface of the bottom of the borehole prior to each application of the ABO. The extent to which rasping actually accelerates the rate of shell penetration by increasing surface area is unknown. Radular activity may also serve in a tactile sense, providing the snail information on the growing size of the break.

The partly dissolved shell fragments scraped from the borehole by the radula are swallowed and can be identified in the stomach of the snail and in its voided fecal strings. To demonstrate the presence of shell material in the stomach, we removed an adult snail from an incomplete borehole on *Mya arenaria* as soon as it completed the rasping period. *Mya* shell was employed because it is soft and the radula removes more of it than of the harder oyster shell (Carriker, 1969). Rasping sounds were heard through a contact microhydrophone. The rasping period lasted about 65 sec. We quickly opened the snail, and in the cecum of the stomach we found white shell raspings bound by mucus into a string-shaped pellet. The pellet was birefringent under polarized light, became red when stained with alizarin sodium monosulfonate, and consisted of distinct solid particles amid soft mushy shell material. The former reacted most strongly under the polarized light and in the stain. Fecal strings collected in small dishes from several snails which were boring small oysters also contained conspicuous white shell raspings which likewise were identified as $CaCO_3$ (Fig. 25). The shell material removed by the radula appears to be dissolved only slightly as it passes down the alimentary canal to be voided in fecal strings and in part resembles the partially dissolved prisms in the incomplete borehole (Carriker, 1969). To what extent $CaCO_3$ from these scrapings is utilized by the snail in its calcium metabolism is not known; it could be determined by the use of isotopes. Fretter and Graham (1962) found that similar shell chips in the gut of the muricid boring snail *Nucella lapillus* were soluble with evolution of CO_2 in acid.

As the diameter of the break at the bottom of the borehole approaches the diameter of the tip of the proboscis, the snail attempts to force the pro-

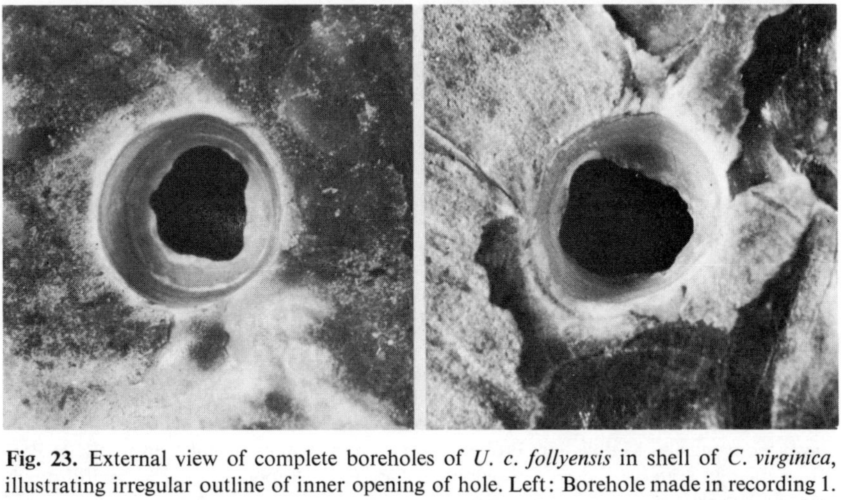

Fig. 23. External view of complete boreholes of *U. c. follyensis* in shell of *C. virginica*, illustrating irregular outline of inner opening of hole. Left: Borehole made in recording 1. Right: Borehole made in recording 2. Outer diameter of boreholes about 1.5 mm. Light photographs.

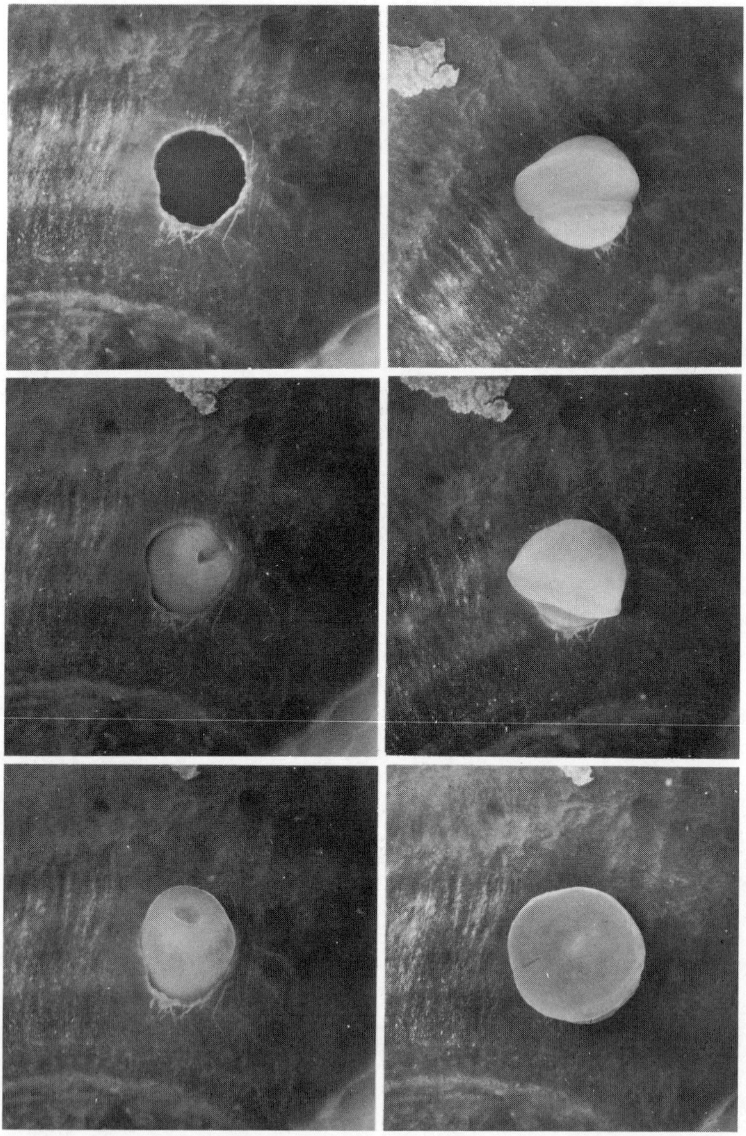

Fig. 24. Final stages of penetration of borehole by *U. c. follyensis* through purple adductor muscle scar in shell of *C. virginica*, valve model. Top left: Empty borehole, organs withdrawn momentarily. Middle left: Proboscis tip examining edge of hole. Bottom left: Proboscis attempting to force itself through hole. Top right: Propodium forcing itself through hole, transverse furrow showing conspicuously, anterior of snail facing the bottom of photograph. Middle right: Propodium moving across (downward) borehole and pedal region anterior to ABO coming into view. Bottom right: Hole enlarged enough to allow ABO to bulge fully through opening. Borehole at top left 0.8 mm in diameter; ABO at bottom right 1.5 mm in diameter. Light photographs.

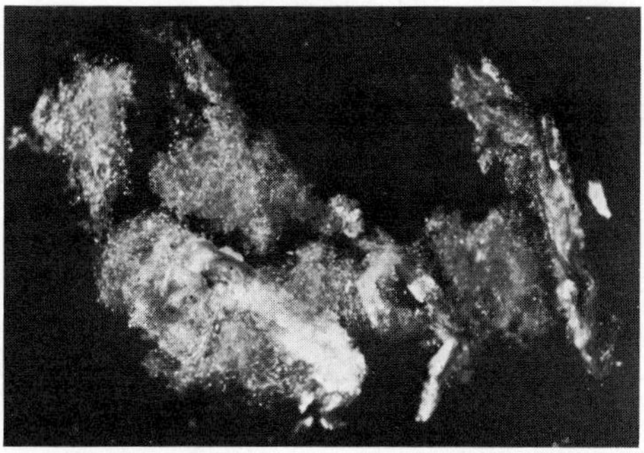

Fig. 25. Fecal pellets of *U. c. follyensis* containing shell raspings from borehole. White areas are shell material which was birefringent under polarized light. Diameter of white pellet about 0.2 mm. Light photograph.

boscis through the opening (Fig. 24. left middle and bottom). This testing is repeated at the beginning of each rasping period, and sometimes at its termination, until the hole is large enough to admit the proboscis. Hole boring is then discontinued, and the snail begins feeding.

In the valve model the emerging hole comes in contact with seawater or air, and the snail is deprived of the stimulus of the mantle tissue on the inside of the valve of the intact prey. If a bit of oyster meat is placed beside the break before it is large enough to accommodate the proboscis, the snail enlarges and constricts its buccal cavity, taking in and discharging minute quantities of seawater (and occasionally air bubbles trapped in the hole) in a chemosensory response, in marked contrast to the normal rasping behavior of the snail. If a fragment of coverglass is placed over the nearly completed borehole, essentially returning the borehole to the incomplete condition, the snail rasps across the glass as if it were the bottom of the borehole, and the cycle of penetration continues for several hours as it did before breakthrough. What prompts the snail eventually to abandon the artifact is not known.

The diameter of the break is enlarged primarily by chemical dissolution, to a minor extent by rasping, and to a lesser degree by propodial activity. As the break widens, the propodium in its turn in the penetration cycle is forced by blood pressure through the opening to such an extent that the sides of the transverse furrow inflate and separate, the furrow resembling exaggerated lips (Fig. 24, top right). As the ledge of the bottom of the borehole thins, pressure from the propodial bulge in some snails often cracks off slivers of shell, which are swallowed during the subsequent rasping period. In other snails this type of propodial activity is negligible, and widening of the break is done by the ABO and the radula. As the primary role of the propodium in hole boring seems to be voidance of seawater from the hole, bulging through the break may be only a functional modification imposed by the presence of the break, and the cracking off of portions of the ledge may be incidental.

As the opening at the bottom of the borehole widens, the ABO is moved gently into the break, so that by time the break is almost large enough to admit the proboscis, the ABO may bulge as much as the diameter of the disc out of the hole (Fig. 24, bottom right). When first inserted, the secretory epithelium pulsates slightly against the edge of the ledge. In the valve model the ABO may bulge into seawater or into air, depending on the position of the model relative to the meniscus. If in seawater, the secretion collects as a thin gel-like film which often remains in position like a translucent milky sheet over the hole when the ABO is withdrawn; if in air, the secretion accumulates as a clear dense syrupy fluid which may pool at the sides of the ABO at its juncture with the shell. For unexplained reasons, the duration of the ABO in position in the borehole after breakthrough varies widely, from a few minutes to an hour and a half or longer. Because the surface of the secretory disc is appressed snugly to the edge of the break, dissolution of the ledge proceeds at a normal rate even underwater. This is easily observed under binocular magnification and its rate can be measured; it seems to vary with the thickness of the ledge. Presumably, dissolution also occurs at a normal rate when snail is boring live prey and mantle fluid washes the emerging ABO.

If a spot of the disc of the ABO extending through the break in a valve model is touched gently with a needle point, this part of the epithelium depresses slightly. If the pressure is increased or is applied suddenly, the ABO will withdraw completely, and the snail will insert the proboscis in the usual manner. By this artifact the chemical phase of the cycle of penetration may be shortened considerably. If the break is enlarged with a needle while the ABO is withdrawn and before the proboscis appears, the radula will rasp and swallow the debris at the edge of the hole and the rate of enlargement of the borehole may thus be accelerated.

The ABO may be protracted at rare times other than in normal boring. In one instance a snail's proboscis tip became securely trapped between the glass and shell of an oyster model, and the snail could not extricate it. The snail then extended its ABO fully and pressed it closely against the shell immediately adjacent to the caught proboscis, as if attempting to dissolve the shell in order to free the proboscis.

So far as we know, all shell-boring gastropods possess a form of the accessory boring organ, and all nonboring gastropods lack it. This is supported by an examination of 29 different species of muricid and naticid borers and many species of nonborers (Carriker, 1961). In 24 species of Muricidae the characteristically shaped ABO is present in the sole of the foot, and in five species of Naticidae it is located under the distal tip of the proboscis.

Recent work by Young (1969) suggests that a shell-boring dorid nudibranch possesses a special gland in the stomodaeum, and an investigation by Day (1969) shows that the shell-dissolving gland in a cymatiid gastropod is located within the proboscis. No predatory gastropods have yet been discovered which bore holes in the shell of prey by mechanical means alone.

The boring behavior of *Eupleura caudata* and *E. c. etterae*, muricid snails closely related to *Urosalpinx*, as observed in oyster models, is identical to that of *U. cinerea*. This is the only other genus of boring gastropods studied in the models to date.

C. Duration of Boring Periods

Because of the difficulty of keeping shucked oysters in good condition within the oyster model for any length of time, and the artificiality imposed by the glass plate and the seepage of oyster attractant and blood through the shell–glass juncture, we could not employ the model to study the normal time sequence of penetration of shell of prey by *U. cinerea*. We circumvented this difficulty by recording radular rasping sounds continuously during the course of complete penetration of normal live oysters. Study of the boring behavior of gastropods by means of rasping sounds has not been attempted before. Kumpf (1964) recorded the sounds produced by the radula of the large nonboring conch *Strombus gigas* while feeding underwater in the field.

1. Recording Rasping Sounds

The microhydrophone, a contact pickup type developed by Barry Martin, consisted of a glass-encased sewing needle and a lead–zirconate–titanate strip connected in turn to a high impedance amplifier and standard two-channel magnetic tape recorder and earphones. The free end of the needle was suspended from a pivoted cantilevered arm and rested on the surface of the oyster adjacent to the boring snail (Fig. 26). Rasping sounds were amplified approximately 1500 times (voltage gain) (Carriker and Martin, 1965).

Rapidly growing, freshly collected young oysters (*C. virginica*) free of encrustations and active adult specimens of *U. c. follyensis* were placed in a large shallow seawater tray next to the recording instruments in a relatively quiet room. Seawater was run very slowly into the tray through a long plastic hose from the laboratory running-seawater system to minimize background noises of the building conducted by the seawater pipes. The microhydrophone was placed on an oyster close to the snail as it started searching for a boring site. This time was designated "zero hour" as a base time for chronological listing of the activities of boring. Recording of sounds originating from both predator and prey was begun on a two-track tape recorder and continued without interruption until penetration of the shell of the prey was completed

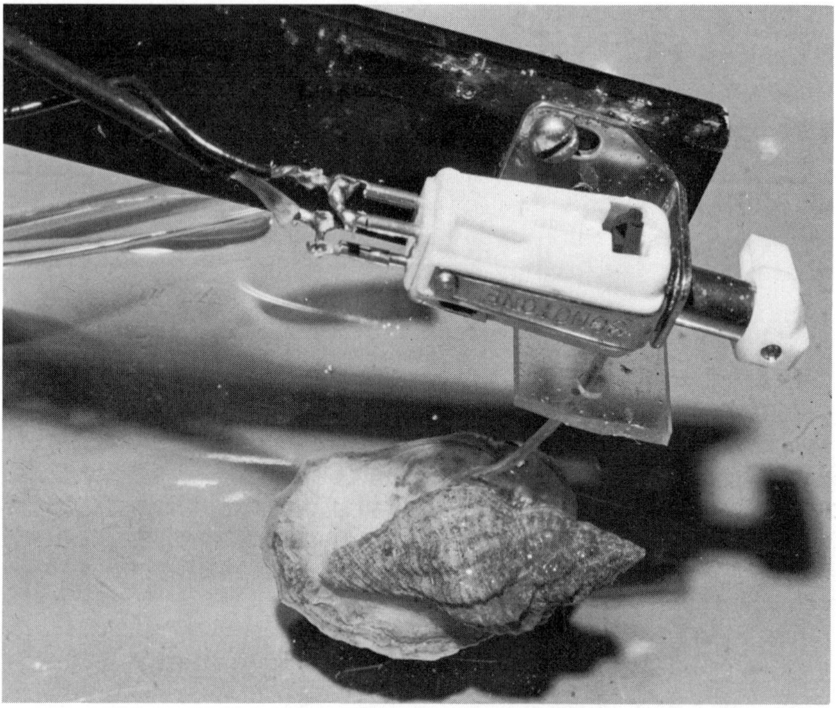

Fig. 26. Contact microhydrophone for recording rasping sounds of *U. c. follyensis.* Snail (38 mm high) is mounted on oyster, *C. virginica;* needle of hydrophone is resting on oyster shell to right of snail. Light photograph.

and the snail was feeding through the borehole on oyster flesh. Two operators alternated day and night watches.

The sounds of rasping, of the snail moving its shell against that of the oyster, and of valve closures by the oyster were heard by the operator through earphones and simultaneously recorded on magnetic tape. Verification of the identity of rasping sounds was made on the oyster model, where rasping could be seen and heard simultaneously; the source of the other sounds was identified by watching the snails and their prey. The operator, with the aid of a stop watch, maintained an additional continuous chronological record on paper of the tape number and side, tape change, time lost in tape changes, type and duration of sounds from the snail and the oyster, number of rasps by the snail per rasping period, and extraneous noises made by the operator and others in the building. These data provided a check on decoding and interpretation of the sounds recorded on the tapes.

During the recordings salinity of the seawater was 32‰, the temperature ranged from 19.0 to 20.4°C, and the light intensity immediately over the snails and oysters was uniformly 0.2 ft-candle day and night. Recording 1 was made September 16–18, 1965. The snail was a female 36.5 mm high. The oyster prey was a fast-growing animal in its second summer and possessed a hard clean shell 28 mm wide and 34 mm long. Recording 2 was made October 11–13 of the same year. The snail, another female, was 37.3 mm high. The oyster, also a fast-growing animal in its second summer, had a clean hard shell and was 19 mm wide and 27 mm long.

The following data were tabulated: (a) time of occurrence and duration in seconds of radular rasping periods, (b) number of individual radular rasps in each rasping period, (c) time of occurrence and duration in minutes of silent inter-rasping periods, which on the basis of studies with the oyster model and the valve model were shown to be predominantly periods of chemical activity by the ABO, (d) time of occurrence and number of movements of the snail shell against the oyster shell, (e) time of occurrence and number of closures of valves by the oysters, and (f) time of occurrence and characteristics of radular rsping sounds at breakthrough when penetration through the shell into the mantle cavity began.

2. Analysis of Duration of Rasping and Dissolution Periods

a. Rasping and Dissolution Periods. Preparations for the recording of penetration of the first oyster were made during the evening of September 16. Shortly after midnight (Eastern Standard Time) a *U. c. follyensis* mounted the upper flat valve of an oyster, and simultaneously we placed other snails on other oysters of the same size as a base of reference to permit visual examination of the progress of penetration by gently lifting the snails off their boreholes. The microhydrophone was balanced on the experimental oyster as soon as the snail was on it, and recording was begun. During the first 96 min from zero time the snail crawled about over most of the surface of the upper valve of the oyster exploring and rasping occasionally for a few seconds at a time. Thereafter, the snail remained in one position directly over the adductor muscle of the oyster and commenced penetration. Breakthrough into the mantle cavity of the oyster was begun about 42.5 hr from time (Fig. 27). This was indicated audibly by the emission of sounds resembling the splintering of shell and the irregularity of the rasping periods. The absence of a record of the duration of rasping sounds at this point on Fig. 27 and earlier at 9 hr was caused by external ambient noises which partly obscured the rasping sounds. After 48 hr the snail was lifted from the oyster; the borehole was completed and the snail was feeding. The borehole was conically shaped in longitudinal section, a characteristic of boreholes by this species of snail in thin shell (Fig. 22, top). The thickness of the oyster shell at the point of penetration, measured after the hole had been sawed in two with a high-speed diamond wheel, was 0.65 mm. The outer circumference of the hole

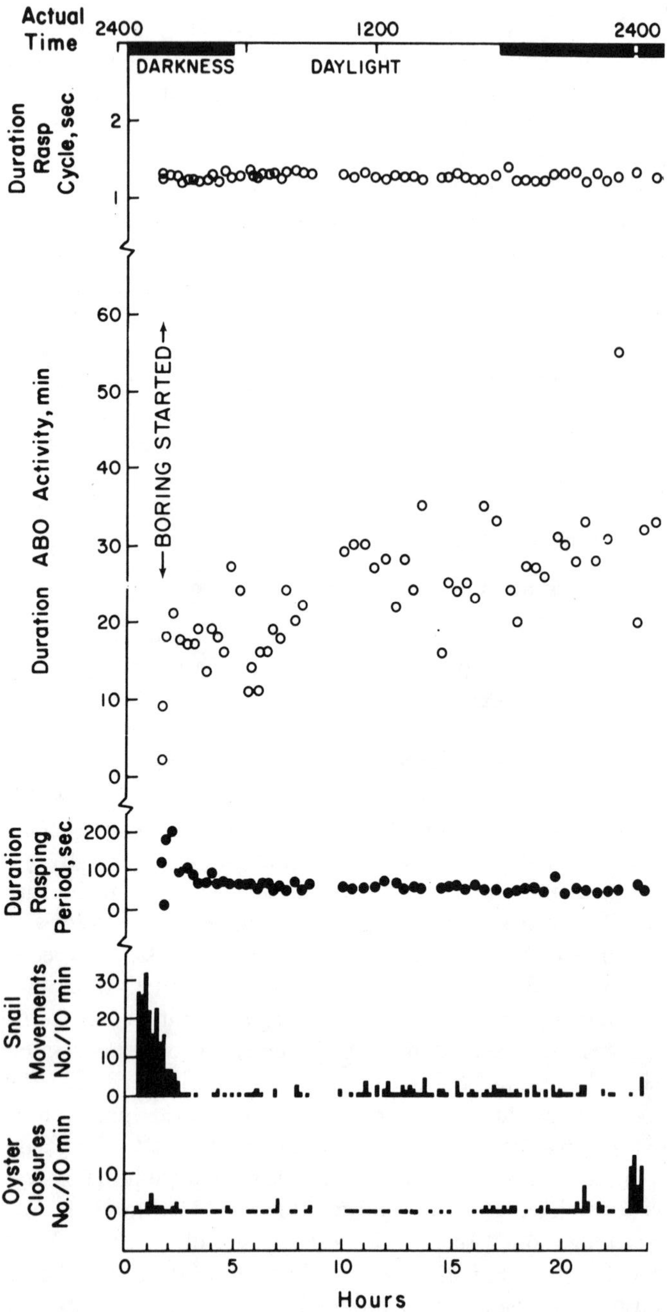

Fig. 27. Duration of rasp cycles, ABO activities, and rasping periods, frequency of contacts of snail shell with oyster shell, and frequency of oyster closures during complete penetra-

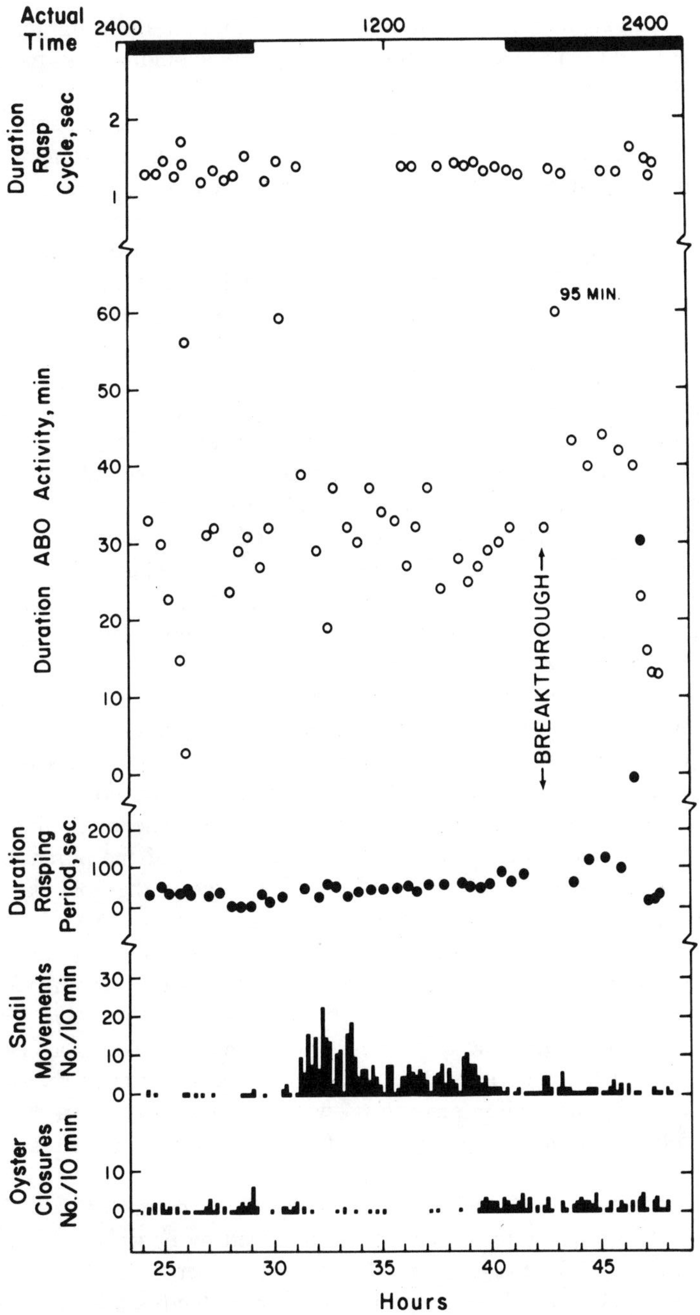

tion of shell of *C. virginica* by *U. c. follyensis*, recorded by contact microhydrophone. Recording 1. (See Fig. 23, left, for photograph of completed borehole.)

was slightly oval, with a maximum diameter of 2.0 mm and a minimum diameter of 1.7 mm. The inner opening was roughly rectangular and measured 1.2 by 1.0 mm. The median section of the borehole revealed that shell in the vicinity of the hole was normally hard and translucent, lacking soft chalky deposits. A characteristic whitish thin layer of shell, resulting from chemical weakening and not removed by the radula, remained around the outer perimeter and the walls of the hole (Fig. 23, left).

The duration of rasping periods in seconds, duration of ABO activity in minutes (inferred from observations on the models), duration of the individual rasp cycle (a mean figure obtained by dividing the duration of the rasping period by the number of rasps for that interval), the number of contacts of the snail shell against the oyster shell, and the number of valve closures by the oyster (both audible through the microhydrophone and confirmed by simultaneous observation of the animals) are plotted in Fig. 27.

Preparations for the recording of penetration of the second oyster were made during the evening of October 11. Shortly after midnight the snail mounted the upper flat valve of the oyster, and we began recording. The snail crawled about over the valve, then off, and finally back onto it, and at 20 min after zero hour began boring in the midanterior portion of the valve. The snail remained in this position until boring was completed at approximately 55.5 hr after zero hour (Fig. 28).

After 62 hr the snail was lifted from the oyster. The borehole was conical in section, and the thickness of the oyster shell at the center of the hole was 0.78 mm. The extreme axes of the outer diameter of the hole measured 1.7 by 1.6 mm, and the diameter of the inner portion of the hole was highly irregular and measured roughly 0.9 mm in diameter. The remaining shelf at the bottom was relatively thin. Shell in the vicinity of the borehole was hard and translucent. Discoloration of the outer edge of the hole was more striking than that on the hole in recording 1 (Fig. 23, right).

The total time for complete excavation of the shell by the snail in recording 1 was approximately 48 hr, an average rate of boring of about 0.33 mm per day. Five to seven hours of this time, as determined in valve models, was consumed enlarging the borehole from the initial break to a diameter which admitted the proboscis. The total time of penetration of the shell to breakthrough by the snail in recording 2 was approximately 55 hr. Allowing another 5 hr for completion of the hole, the average rate of penetration was 0.31 mm per day.

The average rate of shell penetration by the two snails was 0.32 mm per day. Estimates of rate of penetration recorded by Federighi (1931) and Galtsoff *et al.* (1937), ranged from 0.4 to 0.5 mm per day. Since the rate of boring is probably dependent on a variety of factors which have not been examined (such as size of snail, relative quantities of hard and chalky prey shell, temperature and salinity of the seawater, and other unknown factors), these rates are in good agreement with those obtained in recordings 1 and 2.

The sectional shape of both boreholes was a truncated cone. This suggests that the diameter of the secretory disc of the ABO decreased as it extended deeper into the borehole. Whether this decrease in diameter results in a concentration per unit surface area of the secretory cells is not known. If this were so, the amount of secretion per unit area might be expected to increase with depth of borehole, and with it the rate of shell penetration.

As inferred from observations in the models, each period of rasping in recordings 1 and 2 was followed by a period of ABO activity (Fig. 27, 28). After the start of penetration and up to the time of breakthrough, the rasping periods were surprisingly uniform in duration, averaging about 50 sec. As also observed in the models, rasping periods were longer in the early stages of penetration than after boring commenced in earnest. Following breakthrough the duration of rasping periods became widely erratic as the snails rasped off thin edges of the shelf. For example, between hours 46 and 47 in recording 1, rasping periods of 4–14 min were recorded. Minimal durations occurred about three quarters of the distance through the shell between hours 27 and 34.

Duration (arithmetic mean) of individual rasp cycles—that is, extension of the odontophore, rasping, and return of the odontophore into the buccal cavity—in recording 1 was 1.3 sec; minimal and maximal ranges of 1.2–1.7 may be explained by variation in the pressure of rasping and the resistance of the etched substrate (Fig. 27). Average duration of the rasp cycle in recording 2 was 1.4 sec (Fig. 28). A plot of these data demonstrated that the length of the individual rasp cycle did not increase with the duration of the rasping period, suggesting that the rate of muscular activity of the odontophore did not decrease as shell penetration progressed. Likewise, extension of the duration of the period of ABO activity as boring went on did not necessarily lengthen the interval of the rasping period following the longer periods of dissolution. This supports the concept that shell removal is primarily the function of the ABO.

Variation in the duration of each period of ABO activity between start and breakthrough was greater than that of rasping. In recording 1 this ranged from 3 to 59 min. At the beginning of penetration the interval averaged roughly 20 min, and this lengthened gradually to about 30 min as boring progressed. In recording 2 the duration of each ABO period increased to about 22 min during the first 5 hr, remained at this level through hour 33, and then fluctuated widely. The wide fluctuations resembled those in recording 1 after hour 22. Interpretation of the intervals past breakthrough is difficult, as it was not possible to observe how much of the time the snails spent exploring the inner

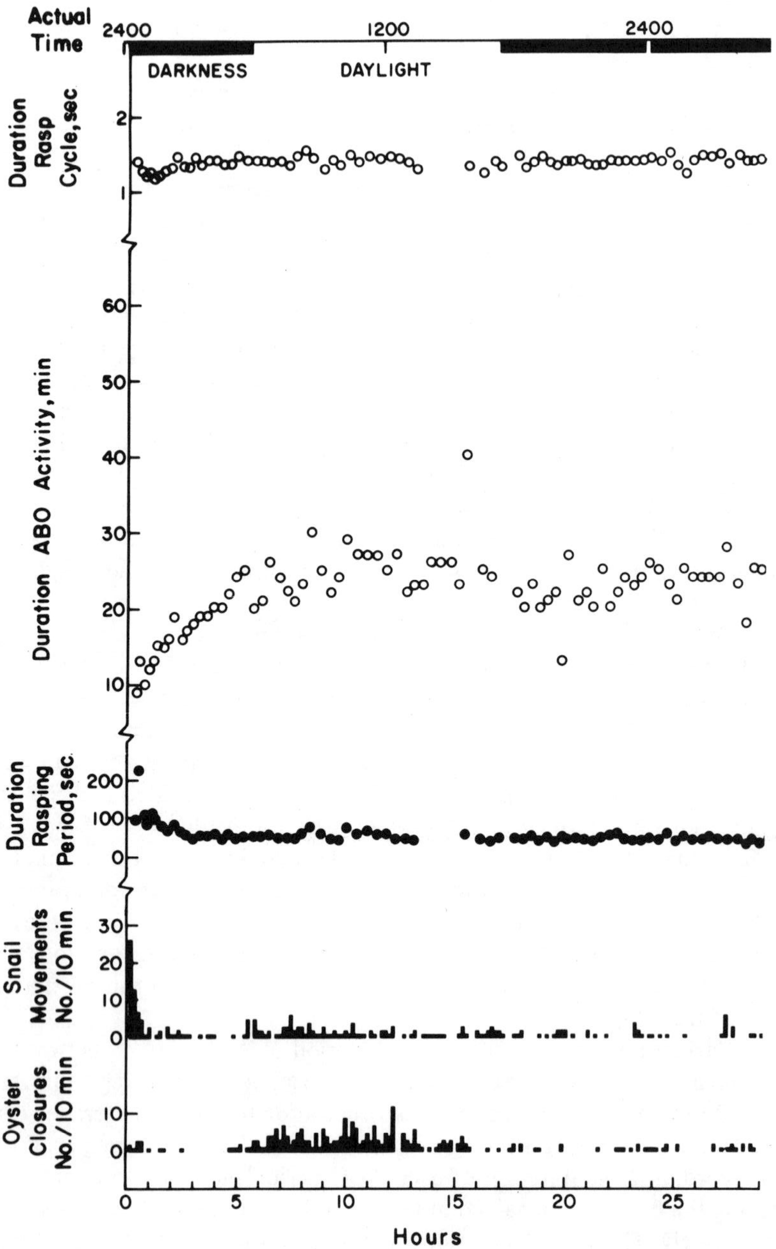

Fig. 28. Duration of rasp cycles, ABO activities, and rasping periods, frequency of contacts of snail shell with oyster shell, and frequency of oyster closures during complete penetration

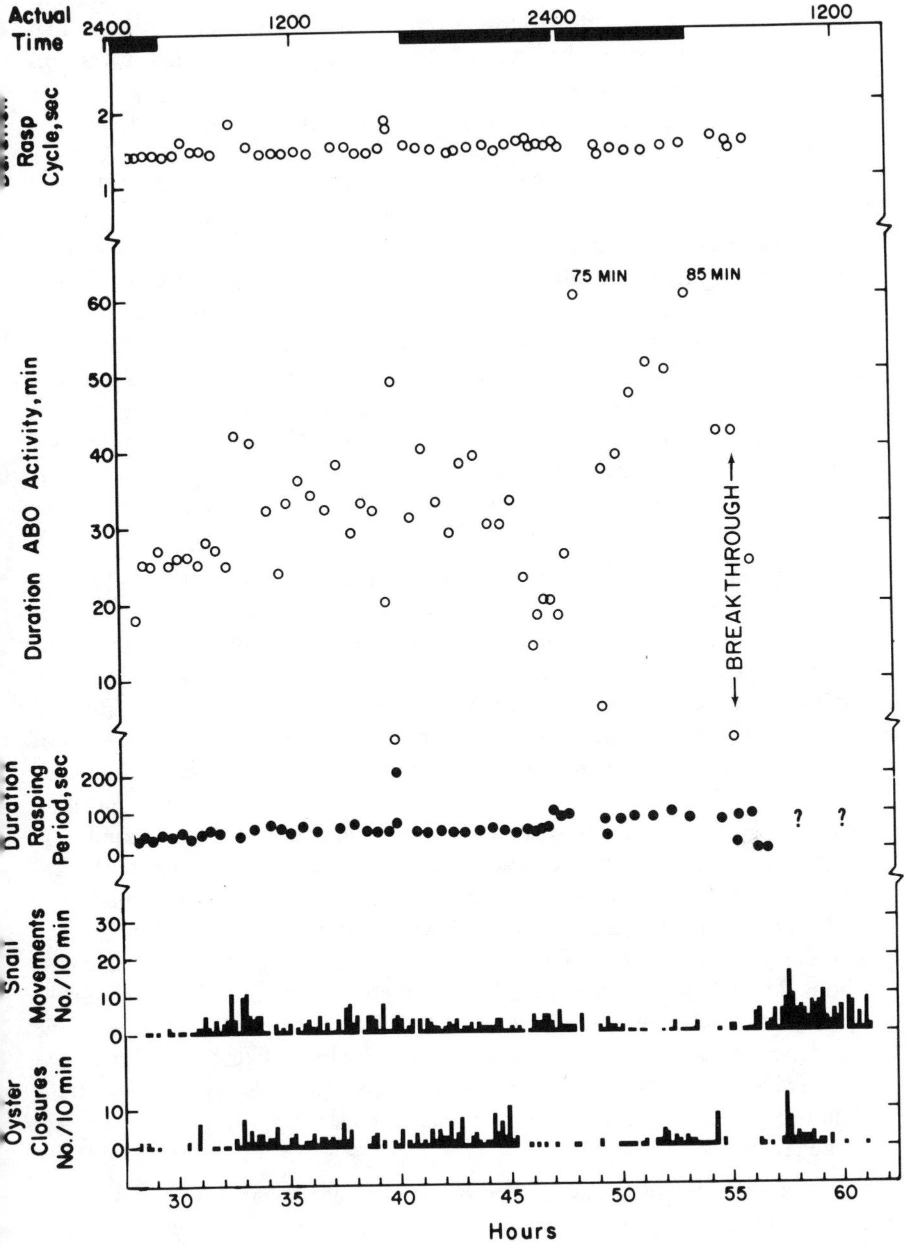

of shell of *C. virginica* by *U. c. follyensis*, recorded by contact microhydrophone. Recording 2. (See Fig. 23, right, for photograph of completed borehole.)

shelf and opening of the borehole with the proboscis and propodium. The fluctuations of length of periods of chemical activity between start and breakthrough are probably accurate representations, as similar variations were observed in the models.

Comparison of the results of recordings 1 and 2, confirmed since by recordings of portions of shell penetration in several additional individuals, demonstrates the similarity of the behavior of boring by the two different snails. Most striking was the uniformity of the duration of rasping periods and rasp cycles. The major difference occurred in the intervals of ABO activity during the first half of penetration. In recording 2 these were considerably less variable than in recording 1. The reasons for this are not clear but may simply reflect individual behavioral differences.

Illumination over the snails in the recording room was purposely maintained uniformly at 0.2 ft-candle throughout the entire period of penetration to obviate the possible effect of excessive or variable illumination on rate of penetration. Under these conditions of dim light, no clear intrinsic effects of the natural outdoor periods of darkness and light were evident on the duration of the rasping periods or other parameters of boring.

b. Contacts of Snail Shell with Oyster Shell. When the foot of the snail is clinging tightly to the substratum, contraction of the columellar muscle brings the shell smartly against the substratum. In recordings 1 and 2 the sound made by the aperture of snail shell striking the oyster valve was clearly audible over the amplifier. The number of shell strikes for each 10 min interval is recorded in histograms in Figs. 27 and 28. In recording 1 the frequency of shell contacts was very high during the initial exploration of the shell surface by the snail. Thereafter, except for a slight increase during the late morning and early evening, the rate was low until the midmorning of the second day, when there was a burst of contacts which began about 11 hr before breakthrough and diminished in intensity as breakthrough was approached. The pattern of shell contacts in recording 2 was similar for the first 48 hr, but the frequency of contacts during the second day was not as great and commenced about 24 hr before breakthrough, and there occurred a third prominent burst of contacts beginning about hour 56. The cause of the variation in the frequency of contacts is not clearly indicated by these data, though increased activity in the vicinity of breakthrough suggests a possible association with it. Additional recordings will have to be made to clear up this interesting aspect of behavior.

c. Oyster Valve Closures. The oysters in recordings 1 and 2 exhibited a wide range of variation in the number of valve closures per unit time after the snail mounted the oysters and started penetration (Figs. 27 and 28). The number of closures varied roughly between 0 and 48 per hour. In recording

1 a slight peak of closures occurred soon after the snail mounted and again between hours 21–29 and 40–48. In recording 2 periods of maximum number of closures occurred between hours 6–16, 31–45, 52–54, and 58–59. Peaks of closures were not associated with time of day. Nor was there any association of concentration of closures with penetration, as the first major mode of closures occurred some 24 hr in recording 1 and 10 hr in recording 2 after initiation of boring. In both oysters, however, the number of closures increased, though erratically, during and after breakthrough, but the number per unit time did not exceed that prior to breakthrough. Futhermore, the modes of valve closures and movements of the snail shell did not coincide consistently in either recording, except after the snails first mounted the oysters, suggesting no response of the oysters to the movements of the snails.

Galtsoff (1964) observed that sudden snapping of the valves of oysters resulting from contraction of the adductor muscle is associated with the discard of rejected food, mucus, detritus, and other particles that accumulate on the inner surface of the mantle and that the frequency of snapping of an undisturbed oyster is relatively uniform and ranges about one to four closures per hour. He reported no periodicity of shell movements of oysters either in running seawater in the laboratory or in the field. He did note, however, that in states of increased excitability such as result from exposure to adverse environmental conditions such as rapid rise in temperature, accumulation of metabolites in stagnant water, and the presence of poisons, the number of closures per hour increased rapidly and irregularly to 20 or more. Galtsoff's observations suggest that the periods of increased number of valve closures by the oysters in recordings 1 and 2 may have been the oysters' responses to the added weight of the snail on their valves, and an "attempt" to dislodge the adverse additional load, or possibly to metabolites released by the predator and identified by the prey as "danger signals." The latter is suggested by the fact that when Galtsoff added extract of the flesh of *U. cinerea* to the edge of the mantle of oysters, the tentacles reacted strongly in 1.5–3.6 sec.

V. RELATIONSHIP OF SIZE OF BOREHOLE, ABO, AND RADULA

A. Borehole Characteristics

Characteristically, the boreholes of *U. cinerea* possess smooth walls, bevelled outer edges, decreasing diameters with depth, a general circularity, and perpendicularity with the surface of the shell of the prey. In very thin shell, boreholes may be parabolic in longitudinal section. With increasing

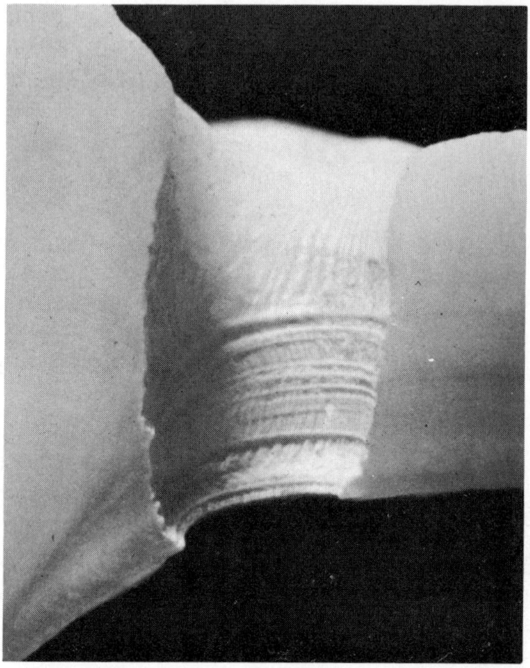

Fig. 29. Longitudinal section of nearly cylindrical borehole excavated by *U. c. follyensis* in aragonitic shell of snail, *Murex fulvescens*. Outer diameter of borehole 1.6 mm. Light photograph.

thickness of shell they assume first a conical shape (Fig. 22) and then a subcylindrical or cylindrical (Fig. 29) shape. Once snails have breached the inner surface of the shell of the prey by an aperture large enough to admit the proboscis, boring is terminated. The shelf which most snails leave partly blocking the inner opening of the hole is characteristically variable in outline. Young snails newly emerged from the egg capsule have the capacity to bore small shelled prey, excavating holes about 0.1 mm in outer diameter (Carriker, 1957); as snails grow in size, the diameter of the hole increases, reaching 2 mm in large adults.

The form of the borehole is dictated by the anatomy of the boring organs, the behavior of boring, and the structure of the calcareous substratum. The radula serves no, or possibly only a minor, part in shaping the borehole. Borehole morphology is the product mainly of dissolution by the ABO secretion; thus the form and size of the borehole generally mirror the shape and diameter of the secretory disc. Decrease in diameter of the borehole with depth results in small part from the fact that the radula, rotated on its axis as it scrapes, grates principally at the bottom of the incomplete borehole, and in large part from the fact that the secretory disc decreases in diameter as the stalk lengthens and the borehole deepens (Nylen *et al.*, 1969; Carriker, 1969; Carriker and Yochelson, 1968). However, the symmetry of boreholes may be so warped by ornamentation, growth irregularities, and differences in hardness of prey shell that perforations may assume a variety of shapes, and exaggerated external shell sculpture may be reflected in the shape of the borehole. Because of the range of variation in the structure of shell, it has not been possible to determine the variation of borehole form produced by an individual snail (Carriker and Yochelson. 1968).

B. Borehole, ABO, Radula, Shell Height, and Other Parameters

Insofar as we could determine in the valve and oyster models, the behavior of boring of individuals of *U. c. follyensis* ranging in shell height from about 15 to 45 mm does not change with increasing shell height. However, whether the size of the borehole increases in proportion to the height of the shell and the width of the radula and whether there are differences in the size of boreholes of male and female snails had not been determined. We investigated these relationships in the following way.

In early December, snails were placed with medium-sized oysters in running seawater warmed to room temperature and were allowed to penetrate for 2–3 days. Boring was then interrupted, and snails and their corresponding boreholes were labeled and measured. Two to four axes of the exterior diameter of each borehole were measured with the ocular micrometer of a dissecting microscope and averaged to give the size of the borehole. The snail was removed from its shell by cracking the shell, and the presence or absence of the ventral pedal gland and the size of the base of the penis relative to that of the right tentacle in the contracted state were noted. A penis approximately half the size of the tentacle, for example, was given a relative size of 0.5, and one the size of the tentacle or larger was designated by 1.0. The width of the radula from outer tip to outer tip of the marginal teeth fully spread was measured in microns with an ocular micrometer in a binocular microscope at the bending plane over the odontophore in the opened buccal cavity.

Eighty-eight snails, 51 females and 37 males, ranging in shell height from 14 to 43 mm were analyzed (Table VI). Identification of males by the presence of the penis was not possible in this subspecies because of the variable size of the vestigial penis in females. In only one female was the penis absent altogether (No, 37. Table VI), and in the remainder it approximated 0.2–0.7 the size of the tentacle. Identification of sex was therefore based on

Table VI. Relationship of Ventral Pedal Gland, Size of Penis, Shell Height, Radular Width, and Size of Borehole in *U. c. follyensis*

Snail No.	Ventral pedal gland	Penis, relative size	Shell height (mm)	Radular width (μ)	Exterior diameter borehole (mm)		
					Maximum	Minimum	Average
			Female snails				
1	Present	0.7	16	176	0.84	0.80	0.82
2	Present	0.5	17	216	0.80	0.73	0.77
3	Present	0.7	19	216	0.90	0.86	0.88
4	Present	0.5	20	235	0.98	0.92	0.95
5	Present	0.5	21	216	0.98	0.90	0.94
6	Present	0.7	21	176	1.02	0.90	0.97
7	Present	0.3	22	176	1.02	0.94	0.98
8	Present	0.7	22	255	0.94	0.90	0.92
9	Present	0.7	24	196	1.06	1.00	1.03
10	Present	0.5	24	235	0.90	0.86	0.88
11	Present	0.3	24	235	1.26	1.14	1.20
12	Present	0.7	24	275	1.18	0.94	1.06
13	Present	0.5	26	255	1.22	1.18	1.20
14	Present	0.7	27	235	1.53	1.33	1.43
15	Present	0.5	28	294	1.18	0.98	1.08
16	Present	0.5	28	314	1.37	1.26	1.31
17	Present	0.7	29	255	0.98	0.90	0.94
18	Present	0.5	29	294	1.37	1.29	1.33
19	Present	0.7	29	275	1.26	1.10	1.18
20	Present	0.7	30	294	1.39	1.20	1.29
21	Present	0.5	31	333	1.31	1.20	1.26
22	Present	0.7	31	275	1.41	1.33	1.37
23	Present	0.7	31	333	1.43	1.24	1.34
24	Present	0.7	32	275	1.73	1.57	1.65
25	Present	0.7	32	314	1.51	1.39	1.45
26	Present	0.5	32	315	1.49	1.37	1.43
27	Present	0.3	32	314	1.40	1.20	1.39
28	Present	0.5	33	314	1.18	1.18	1.18
29	Present	0.7	33	314	1.57	1.49	1.53
30	Present	0.3	34	314	1.84	1.69	1.77
31	Present	0.7	34	216	1.47	1.53	1.50
32	Present	0.2	34	235	1.49	1.41	1.45
33	Present	0.3	34	333	1.65	1.49	1.57
34	Present	0.5	35	343	1.55	1.43	1.49
35	Present	0.5	35	294	1.80	1.69	1.75
36	Present	0.5	35	333	1.65	1.33	1.49
37	Present	0	36	373	1.69	1.53	1.61
38	Present	0.3	37	373	1.68	1.37	1.50
39	Present	0.7	37	333	1.88	1.69	1.79
40	Present	0.3	37	353	1.69	1.65	1.67
41	Present	0.5	38	373	1.49	1.41	1.45
42	Present	0.6	38	333	1.33	1.22	1.27
43	Present	0.2	38	353	1.61	1.41	1.51
44	Present	0.5	38	353	1.12	1.10	1.11
45	Present	0.2	39	353	1.39	1.33	1.36

Table VI. Cont'd

Snail No.	Ventral pedal gland	Penis, relative size	Shell height (mm)	Radular width (μ)	Exterior diameter borehole (mm)		
					Maximum	Minimum	Average
46	Present	0.3	39	353	1.10	1.02	1.06
47	Present	0.3	39	353	1.80	1.57	1.70
48	Present	0.3	40	353	1.65	1.41	1.53
49	Present	0.3	41	373	1.65	1.61	1.63
50	Present	0.2	41	373	2.04	1.77	1.89
51	Present	0.2	42	353	1.61	1.45	1.53
				Male snails			
1	Absent	1.0	14	167	0.49	0.43	0.46
2	Absent	1.0	15	176	0.71	0.65	0.68
3	Absent	0.7	18	216	0.82	0.75	0.79
4	Absent	1.0	18	176	0.78	0.71	0.75
5	Absent	1.0	18	226	1.39	1.24	1.32
6	Absent	1.0	19	216	0.94	0.86	0.90
7	Absent	1.0	19	196	0.94	0.86	0.90
8	Absent	1.0	19	216	0.98	0.92	0.95
9	Absent	0.7	20	216	0.98	0.90	0.94
10	Absent	1.0	20	235	0.92	0.88	0.90
11	Absent	0.7	20	216	1.08	1.04	1.06
12	Absent	0.5	21	216	0.82	0.78	0.80
13	Absent	0.5	21	235	0.98	0.90	0.94
14	Absent	1.0	22	235	0.88	0.82	0.85
15	Absent	1.0	22	216	0.90	0.82	0.86
16	Absent	1.0	23	216	1.06	0.90	0.98
17	Absent	1.0	23	255	1.10	0.96	1.03
18	Absent	1.0	23	255	1.10	1.06	1.08
19	Absent	1.0	23	235	1.02	0.92	0.97
20	Absent	1.0	23	255	0.86	0.78	0.82
21	Absent	1.0	25	255	0.98	0.86	0.90
22	Absent	1.0	25	255	1.10	0.92	1.01
23	Absent	1.0	25	255	0.78	0.71	0.75
24	Absent	1.0	26	255	1.22	1.02	1.10
25	Absent	1.0	26	255	1.22	1.14	1.17
26	Absent	0.6	27	255	1.22	1.06	1.14
27	Absent	1.0	27	294	1.18	1.10	1.15
28	Absent	1.0	27	245	1.33	1.20	1.26
29	Absent	1.0	28	275	1.35	1.18	1.26
30	Absent	0.7	28	294	1.37	1.29	1.33
31	Absent	1.0	29	294	1.37	1.29	1.33
32	Absent	1.0	30	255	1.28	1.18	1.23
33	Absent	1.0	31	333	1.65	1.45	1.55
34	Absent	1.0	33	294	1.39	1.31	1.35
35	Absent	1.0	33	294	1.41	1.29	1.34
36	Absent	1.0	35	333	1.65	1.29	1.35
37	Absent	1.0	39	333	1.65	1.53	1.59

the presence or absence of the ventral pedal gland, which shapes the egg capsule and is present only in females. The size of the live ABO varies significantly, depending upon the degree of relaxation. Numerous attempts to obtain an accurate index of the normal size of the excised ABO by means of chemical relaxers proved futile, so we accepted the average diameter of the exterior opening of the borehole as a measure of the size of the normal functional ABO.

The shape of the borehole excavated by both sexes in oyster shell is generally slightly oval (Table VI). Examination of the outline of the hole relative to the orientation of the foot of 20 snails indicated that the longer axis of the hole ranged from parallel to 90° (most around 30°) to the anterior–posterior axis of the foot. Study of relaxed excised ABOs showed that the secretory disc is likewise oval, the ratio of the longer to the shorter axis approximating that of the two axes of boreholes. This anatomical correlation of the secretory disc of the ABO and the borehole is further evidence for the major role of the ABO in shaping the borehole. Why the long axis of the borehole in different individuals should vary relative to the longitudinal axis of the snail's foot is not known.

This study disclosed that within the range of sizes recorded in Table VI increase in shell height in both sexes was accompanied by a proportionate increase in radular width (Fig. 32). Similar relationships were found for shell height and diameter of borehole (Fig. 30) and radular width and diameter of borehole (Fig. 31). With two exceptions, the width of the radula of different snails of a given shell height did not vary more than about 75 μ (Fig. 32). The exceptions, two female snails of a shell height of 34 mm with radulae 216 and 235 μ wide, respectively (snails 31 and 32), or a width about 100 μ below the normal, may have been regenerating their proboscides; this suggestion

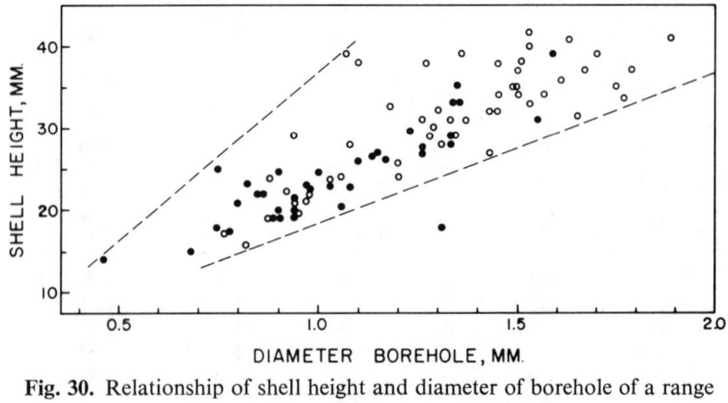

Fig. 30. Relationship of shell height and diameter of borehole of a range of sizes of *U. c. follyensis* (see Table VI).

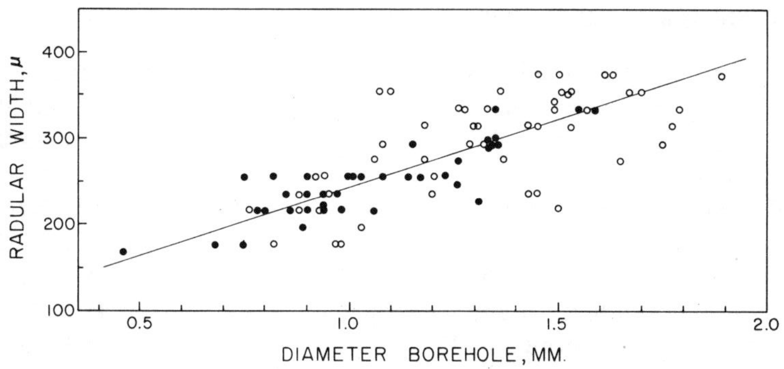

Fig. 31. Relationship of radular width and diameter of borehole in range of sizes of *U. c. follyensis* (see Table VI).

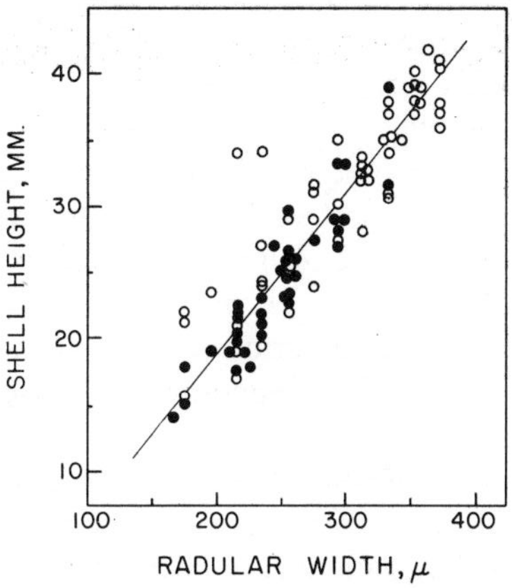

Fig. 32. Relationship of shell height and radular width in range of sizes of *U. c. follyensis* (see Table VI).

is supported by the fact that the size of their boreholes fell within the average diameter for their shell heights. The diameter of boreholes, on the other hand, varied widely among snails of a given shell height and of a given radular width (Figs. 30 and 31), this range being as much as 0.3 mm in small snails and 0.8 mm in the larger individuals. This variation is a reflection of the wide range in size of the ABOs in snails of the same size group and further evidence for the negligible part played by the radula in shaping the borehole. Only rarely do snails bore disproportionately large holes: such is the case of male snail 5 (shell height 18 mm), which excavated a borehole 1.32 mm in diameter (see point below dotted line, Fig. 30).

C. Variation in Size of Boreholes

Discovery of the broad variation in the diameter of boreholes excavated by different snails in the same height range (Fig. 30) raised a question about the extent of this variation in boreholes made by the same snail in a relatively short period.

We pursued the problem as follows. Individuals of *U. c. follyensis* ranging in shell height from 16.0 to 40.5 mm (see Table VII) were placed in a running seawater tray with prey: snails 1–4, 7, and 8 were isolated in January–March with *C. virginica* ranging in diameter from 1 to 3 cm in seawater which was warmed to 17–21°C in a heat exchanger and degassed by aeration; snails 5 and 6 were isolated in May–August with small oysters and *Mytilus edulis* in seawater at ambient temperature which increased during the summer from approximately 16 to 24°C. Salinities throughout the study ranged from 30.5 to 31.3 ‰. Snails bored at random, and in two thirds of the borings they were allowed to feed their fill prior to the next boring. In the remaining borings the prey were withdrawn before completion of the prior borehole to note the effect of absence of feeding on the size of the borehole. The external diameter of the borehole was measured, four axes for smooth holes and as many as eight axes for the more irregular holes.

The change with time of the average diameter of boreholes of each of snails 1–8 is plotted in Fig. 33. The changes in height of the shells of these snails during the period of observation and of maximum diameter (both decrease and increase) of the corresponding successive boreholes are tabulated in Table VII.

The height of the shell of snails 6 and 8 increased 0.8 and 0.2 mm, respectively; that of snails 1, 2, and 4 remained constant; and that of snails 3, 5, and 7 decreased 0.1, 0.2, and 0.2, respectively. Erosion of the spire and siphon tips of the shell accounted for the decrease. Six empty snail shells in the same height range were placed in running seawater during the period of the observations as controls. The height of four of these remained constant, and that of two decreased 0.1 mm.

The experiment demonstrated that boreholes of individual snails vary widely in size in a short period of time (Fig. 33, Table VII). The average diameter of the boreholes of snails 1, 3, 6, and 7 increased gradually, whereas the diameter of snails 2, 4, and 8 fluctuated conspicuously with time. Maximum increases and decreases in size of boreholes were demonstrated by snail 5, with a net decrease of 17 % during the course of the observations, possibly suggesting that the snail was in ill health. Although the diameter of boreholes increased with shell height of snails (Fig. 30) and thus undoubtedly with overall consumption of food, individual meals did not always result in an increase in diameter of the subsequent borehole in this experiment

Table VII. Variation in Size of Boreholes of Individual *U. c. follyensis* in 1–2 Month Period[a]

Snail	Snail height (mm)		Change in diameter of successive boreholes: Maximum increase		Change in diameter of successive boreholes: Maximum decrease		Size of borehole at start and finish of observation	
No.	Start	Finish	mm	%	mm	%	Diameter (mm)	Percent change[b]
1	17.2	17.2	0.11	14	0.05	5	0.74–1.02	+38
2	16.0	16.0	0.32	41	0.18	16	0.78–0.86	+10
3	28.7	28.6	0.12	11	0.06	4	1.10–1.36	+24
4	25.2	25.2	0.11	10	0.14	12	1.19–1.24	+4
5	27.2	27.0	0.40	41	0.40	30	1.27–1.06	−17
6	35.8	36.6	0.12	9	0.03	2	1.38–1.46	+6
7	36.6	36.4	0.11	8	0.07	5	1.38–1.42	+3
8	40.5	40.7	0.33	23	0.19	11	1.42–1.62	+14

[a] Snails No. 1–4, 7, and 8 in winter; snails No. 5 and 6 in summer.
[b] Decrease (−); increase (+).

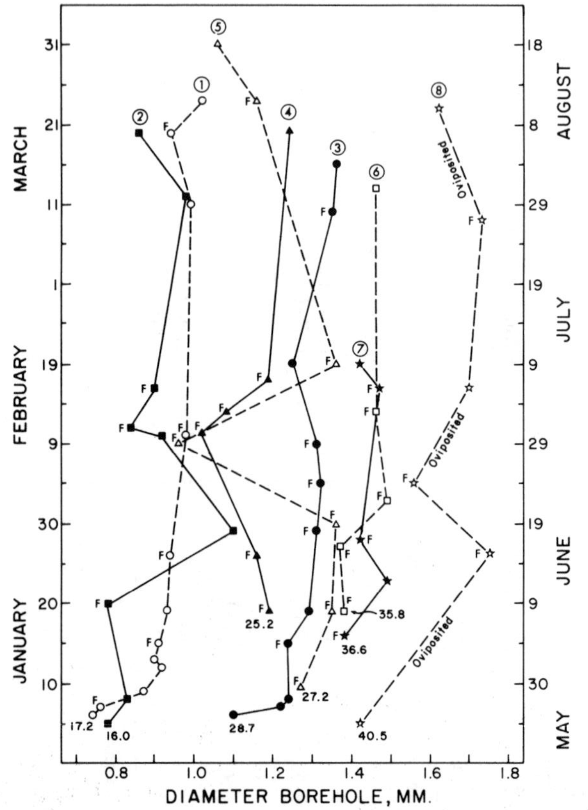

Fig. 33. Change of size of boreholes of eight individual *U. c. follyensis* with time. F = snails permitted to feed after boring hole to note effect of feeding on subsequent borehole. At points without the F snails were not allowed to feed.

(Fig. 33). In fact, in several instances just the reverse occurred. This apparent inconsistency suggests that the size of the ABO upon extension during any particular hole boring may depend upon the health of the snail, blood pressure in particular, or upon other physiological parameters whose effect upon the size of the extended ABO is unknown at present. These observations provide further proof of the negligible role of the radula in determining the size of the borehole.

D. Maximum Depth of Boreholes

How deeply into an incomplete borehole the stalk permits extension of the secretory disc of the ABO, and thus what maximum thickness of shell can be perforated by a snail of a given size, has not been reported. There is also the perennial question of whether snails avoid portions of shell which are too thick for complete penetration and seek thinner ones which they can penetrate.

To seek an answer to these questions, we placed a large population of adult *U. c. follyensis* with 15 large (9.5–15.5 cm long) oysters (*C. virginica*) in running seawater at room temperature and left them together for a month, during which time the oysters were bored, consumed, and subsequently abandoned. The external diameter and depth of the resulting boreholes were then measured. Boreholes with depths exceeding 1.5 mm are plotted in Fig. 34 (the mean shell height of the snail excavating each hole can be approximated from Fig. 30).

Deepest incomplete boreholes ranged in depth from 4.3 to 5.3 mm, and their corresponding diameters varied from 1.16 to 1.65 mm (shell heights of the snails boring these holes varied from 30 to 45 mm). The gentle increase in depth with increasing diameter of the deepest incomplete boreholes (zone between dotted and solid lines) suggests that these depths may approach the maximum penetrability for snails this size. This suggestion is supported by the absence of complete boreholes in this zone (see plot of complete boreholes, Fig. 34). The scatter of incomplete boreholes below the dotted line resulted mainly from snails abandoning boreholes when prey gaped after being penetrated at relatively thin spots in the shell by other snails.

These studies confirm our observations and those of others that snails seek a boring site on the shell of prey without regard to thickness of the shell and suggest that in extremely large prey with thick shell they are not able to complete perforations. In this event, the incomplete excavation is abandoned. The length of the stalk probably varies among snails of a given size, accounting for the variability of maximum depth of incomplete boreholes for a given size borehole between the solid and dotted lines, but it has not been possible to determine this.

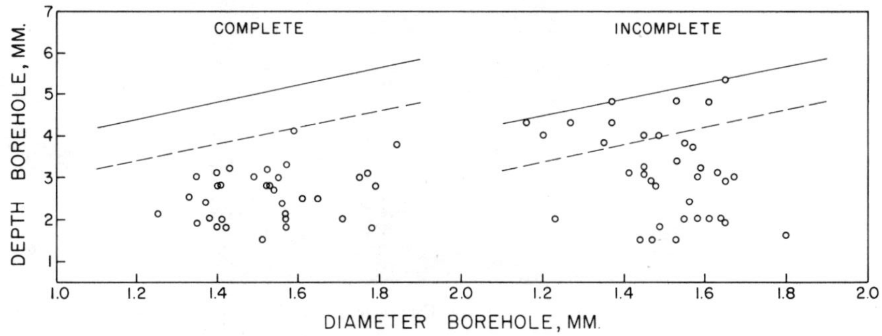

Fig. 34. Depth and diameter of complete and incomplete boreholes of *U. c. folleynsis.*

VI. FEEDING

A. Penetration of Mantle Cavity of Prey

As soon as the inner opening of the borehole is large enough, the snail extends its proboscis through it to feed on the flesh of its prey.

In a valve model from which all flesh has been removed, the proboscis upon emergence from the borehole encounters seawater, rather than mantle, and, conditioned to expect food, explores the valve a maximal radius approaching the length of the proboscis. In the course of the search the mouth and buccal cavity dilate frequently, taking in small quantities of seawater in "tasting" movements. The intake of seawater was confirmed by watching (a) the movements of minute particles of detritus suspended naturally in the seawater and (b) puffs of a dilute vital stain introduced in the proximity of the mouth. After a period of fruitless seeking, which may last more than

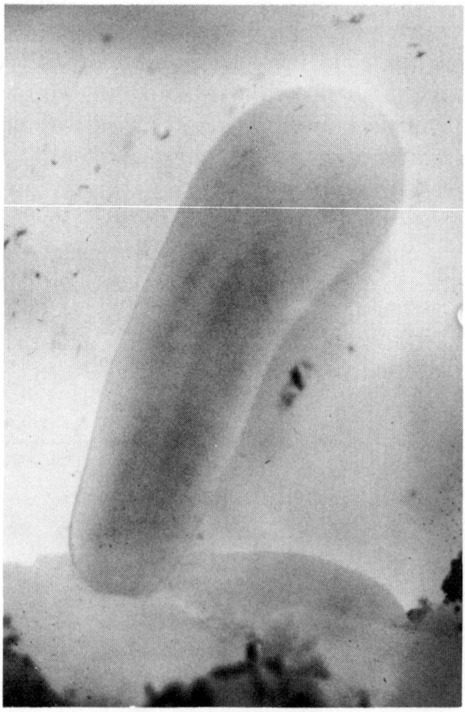

Fig. 35. Proboscis of *U. c. follyensis* feeding through borehole (upper right) on adductor muscle (white tissue at bottom) adhering to valve of *C. virginica* in valve model. Diameter of middle of exposed proboscis about 1.5 mm. Light photograph.

a half hour, the snail withdraws its proboscis from the hole and abandons the model.

The capacity of snails to detect and locate fresh flesh is well developed. Muscle adhering to the adductor muscle scar in the valve model was readily located by the proboscis and fed on so long as the distance from the borehole to the flesh was not greater than the length of the proboscis (Fig. 35). After an unsuccessful search for food on a valve devoid of flesh and after withdrawal of the proboscis from the borehole, snails reinserted the proboscis through the borehole to feed on fragments of oyster flesh placed by us beside the borehole.

After excavating through the shell of an intact live oyster, the snail encounters the mantle and commences to feed on it immediately. A perforation is soon eaten through the mantle, and the proboscis is extended into the mantle cavity. Choice of organs then appears to be more or less random, the proboscis going from one tissue to another in an irregularly expanding radius limited by the length of the proboscis. To what extent a snail exhibits a preference for particular organs is not known. Although there is a tendency in feeding to favor the softer tissues, this is not exclusively so, and the adductor muscle is also consumed. The borehole, for example, frequently is made directly over the adductor muscle, and upon penetration of the shell the snail eats an opening through this into the mantle cavity.

B. Biting and Swallowing

Food particles are torn free from the flesh of the prey by the radula. The food-grasping stroke is essentially the same as that employed in rasping shell; the main difference is that caused by variation in the solidity of the rasped surface, which places the maximal stress on the odontophoral muscles at different phases of the stroke. The shell-rasping stroke consists of the protracting phase which brings the radula forward and ventrad against the surface of the shell, the rasping phase in which the teeth are scraped against the shell, and the retracting phase which conveys bits of dislodged shell material to the esophagus. In the boring operation the radular muscles expend maximum energy as the teeth are scraped against the shell. The retracting phase requires but little energy because of the light load of shell particles carried on the radula into the buccal cavity.

When flesh is grasped, on the other hand, the greatest expenditure of energy comes during the initial part of the retracting phase, when flesh is torn loose. In biting, the snail pushes the odontophore against the flesh, and as it slides the radula over the cartilages the teeth are thrust into the flesh. The rachidian and marginal teeth come erect at the bending plane, and as the marginal teeth begin to fold over the rachidian teeth in their passage into the odontophoral sulcus an effective five-cusped grasping or biting organ is formed by each transverse row of teeth which grips the flesh (Fig. 11). At this instant, the powerful odontophoral retractor muscles contract and the morsel of food is torn loose and, impaled on the posteriorly pointing teeth, is conveyed into the buccal cavity. The whole proboscis responds to the impact of this retraction (Carriker, 1943).

Removal of food from the radula in the buccal cavity appears to take place primarily by suction. In its posterior passage into the buccal cavity, the radula overfolds into the sulcus and passes immediately under the sclerite and the dorsal carina, which border the opening of the esophagus. Simultaneously, the numerous esophageal tensors contract and the anterior part of the esophagus is enlarged, creating a difference in hydrostatic pressure between the interior and exterior of the buccal cavity. Synchronously, the upper surface of the odontophore is applied to the esophageal opening, and food particles are sucked off the posteriorly pointed teeth and carried into the esophagus. By ciliary and peristaltic activity the particles are transported

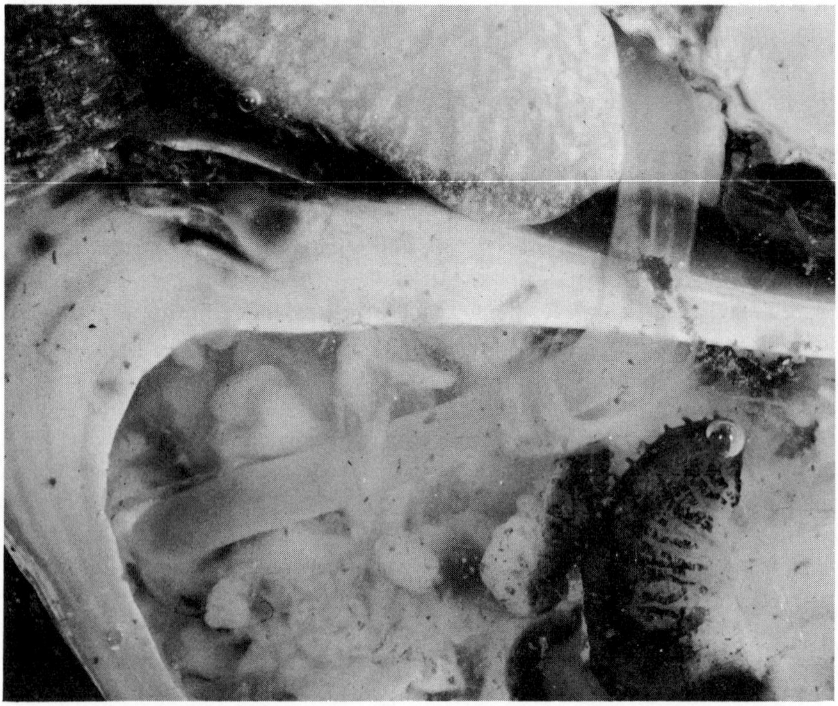

Fig. 36. Proboscis of *U. c. follyensis* inserted between shell and glass into cavity of oyster model feeding on flesh of *C. virginica*. Anterior of snail facing top right of the photograph. Diameter of terminal portion of proboscis 1.5 mm. Light photograph.

to the stomach. Food particles are whipped with surprising speed into the esophageal bulb after each retraction of the odontophore. The thinness of the dorsal and ventral esophageal grooves and the translucency of the proboscidial integument facilitate observations of swallowing (Fig. 36). In a snail with the proboscis fully extended, at a temperature of 28°C, food particles passed down the esophagus at an average rate of 2 mm/sec. This speed varied slightly with the size of the food particle. Ingestion of loose food material such as mucus and ova is done mostly through the sucking movements of the buccal cavity and the esophageal bulb, while the radula and associated structures remain at rest. Any objectionable particle once in the buccal cavity or the esophagus is promptly ejected by a reversal of the functions of the esophageal bulb and buccal cavity (Carriker, 1943).

C. Rate of Feeding

Individuals of *U. cinerea* have been observed to feed continuously on exposed oyster flesh in the laboratory for as long as 2 days at a time. The snails literally bury themselves in the flesh as they feed. Under natural conditions they feed uninterruptedly on large intact live oysters for longer periods. The number of prey destroyed by these snails per unit time increases as the size of the prey decreases, and large snails destroy more prey than do small snails. Excessive exposure to air, as intertidally, curtails boring, and boring rate increases during the breeding season. The maximum average number of small oysters destroyed by small snails is recorded as 34 per week per snail. On the average, adult snails destroy oysters 4–6 cm long at the rate of about 0.14–0.35 per week (Carriker, 1955*a*; Hancock, 1959). Maximal rates, however, may be considerably higher. Galtsoff *et al.*, (1937) for example watched a single snail penetrate and consume an oyster 5 cm long in 24 hr. Wood (1965) observed that at Nobska Point, Massachusetts, where the barnacle *Balanus balanoides* has been dominant for a long time, an adult barnacle may be approached, penetrated between opercular plates, and cleaned out by a native *U. cinerea* within 20 min. He further noted that inexperienced predators upon barnacles drill a hole through or between compartmental plates and require several hours.

In a bifactorial study of the combined effects of salinity and temperature on *U. cinerea* from Long Island Sound, Manzi (1970) found that, at all temperatures studied, a salinity of 12.5 ‰ was near the lower limit for feeding on young oysters (3–13 months old), that feeding rates increased with increase in both temperature and salinity, and that maximum feeding rates, 1.16 oysters per week, were observed in 25°C and 26.5 ‰, the highest temperature–salinity combination studied. Hanks (1957) noted that the lower limit of temperature for feeding was about 7.5°C, though feeding at this temperature was intermittent, and that feeding rate increased steadily as the temperature of the water was raised to 25.0°C, but decreased as the temperature was increased to 30°C.

D. Gaping of Oysters

The valves of an oyster remain tightly closed for an undetermined time after a snail penetrates and begins to feed. Presumably, consumption of some organs, particularly the adductor muscle, results in gaping of the valves sooner than when other organs, such as the mantle, are attacked. There is no information on this subject, nor is it known to what extent infections resulting from microorganisms in seawater, pseudofeces, and feces in the mantle cavity produce infections which might hasten mortality. Federighi (1931) performed experiments which suggested to him that *U. cinerea* while perforating an oyster secretes a toxic substance which kills it. His provocative observations have never been confirmed. It would be necessary to prove that death resulted from a toxic chemical (possibly the secretion of the ABO or secretions from the buccal cavity, or both) and not from infection and bleeding resulting from the tearing of flesh by the radula.

As feeding by the snail(s) continues, the adductor muscle of the oyster weakens and relaxes, and the valves begin to gape slightly. At this point, other snails in the vicinity, attracted by the fluids from the gaping oyster, approach the moribund prey and extend their proboscides between the valves to join in the meal (Fig. 37). In the laboratory where we maintain *U. c. follyensis* in dense concetrations, it is not uncommon to find snails aggregated around the edges of the valves of such oysters. Feeding between the valves of gaping oysters in this manner is possible because the adductor muscle has become inactivated and the valves are immobile, and danger of proboscides being clamped between them has passed. According to Pope (1910–1911), it is also common for two snails to extend their proboscides through a common borehole and feed side by side.

As more of the oyster is consumed, its valves spread more widely apart from the pull of the ligament, and snails then crawl inside onto the flesh to feed. The snail(s) responsible for the initial killing invariably continues to feed through its borehole. Small crabs and fish are also attracted to the wounded oyster and likewise join in its consumption. This stage of feeding, however, can be hazardous to the snails, as crabs and fish may occasionally bite off a proboscis embedded deeply in the oyster flesh. How frequently this misadventure occurs in nature has not been determined. That it does take place from time to time is confirmed by the fact that we have dissected adult snails which contained proboscides the anterior portion of which was abnormally small and resembled those which were regenerating after experimental amputation. The remarkable capacity of this species to regenerate

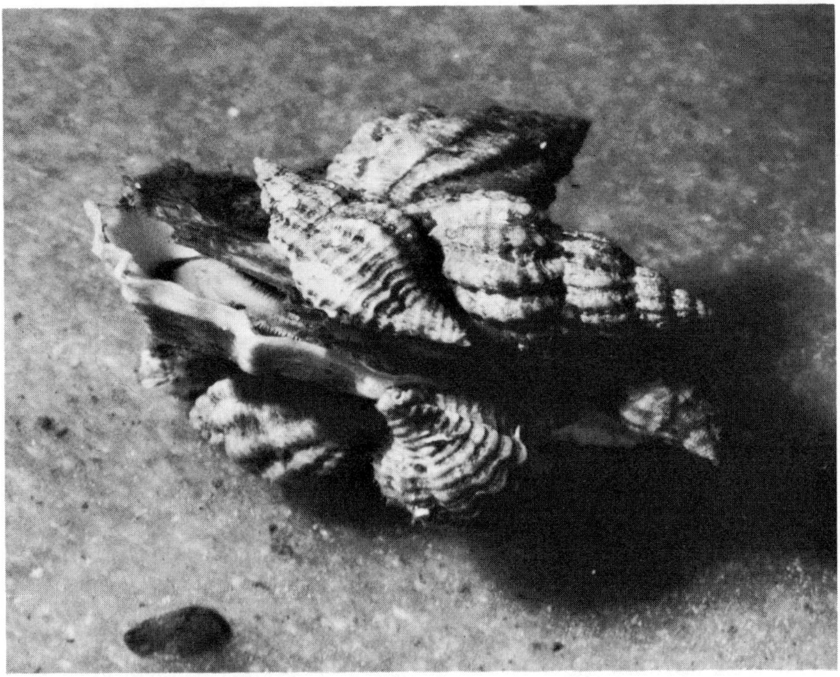

Fig. 37. *U. c. follyensis* individuals, attracted by fluids from gaping oyster which is being bored by other snails, are feeding on oyster between the valves. Largest snails 40 mm high. Light photograph.

amputated proboscides in a relatively short time is perhaps nature's way of coping with the problem and protecting the species (Carriker, 1961).

VII. DISCUSSION AND CONCLUSIONS

This chapter presents an analysis of the behavior of the appetitive and consummatory phases of the feeding activity of the adult predatory marine gastropod *Urosalpinx cinerea*. The investigation was made with reference to the anatomy and function of the organs involved and in relatively constant environmental conditions. The analysis suggests that the detection, approach, penetration, and ingestion of shelled prey by this species is a remarkable example of complex behavioral specialization. As stressed in the following discussion, however, there still exist major voids in our knowledge of the predatory behavior.

The chemical signals which attract *U. cinerea* to oysters has been identified by other investigators only as one or more of the metabolic end-products of the prey (Blake, 1960; Wood, 1968). The attractant has not yet been characterized. Successful synthesis of the substance could provide a bait for trapping the snails on oyster beds. Successful control of this serious predator would be a boon to shellfishermen (Carriker, 1955*a*). The organ in the oyster which produces the attractant has not yet been identified. The kidney is one logical source. Another may be the mat of epidermal secretory cells which covers the free surfaces of the oyster (Galtsoff, 1964). Other than Janowitz' preliminary observation (Blake, 1960), nothing has been reported about the effect of kind and quantity of food consumed by the prey or the nature of the attractant. The specific role of the osphradium, tentacles, propodium, mantle edge, proboscis, and other external surfaces of the snail in distance and close-range detection of the attractant is still uncertain (Kohn, 1961). Our behavioral observations suggest that the osphradium is used in distance detection, the propodium and tentacles in close-range detection, and the propodium and proboscis tip in contact or very close-range sensing. Electrophysiological methods would help in determining the receptor sensitivity of each of these organs. The effect of environmental factors in the sea on attenuation and destruction of the attractant has not been investigated. It should be demonstrated, for example, to what extent decreasing temperatures in the fall and decreasing salinities up estuaries alter its attractiveness and whether microorganisms, suspended silt and clay particles, and natural organic substances decrease its potency. Whether the attractant adsorbs naturally to the surface of oyster valves has not been reported. Our initiatory experiment was inconclusive. Should adsorption occur, it would be important with reference to the response of snails to oysters to know whether the attractant is longer-lived adsorbed or in solution in seawater. It appears from our initiatory experiment that snails may respond to substances released by what may be distinctive sessile organisms attracted to the surface of oyster shell by the effluent from the mantle cavity. The significant ecological implications of this possiblility in oyster culture should prompt further research.

We hypothesized that dissolution of $CaCO_3$ and organic matrix at the valve surface during normal erosion of the shell might aid snails in recognition of prey and conjectured that the external form of the valves of prey might also facilitate recognition. Our preliminary observations, however, were not informative. To solve these problems, it will be necessary to seal the attractant within live oysters for several days at a time without injuring them. We explored a variety of artificial sealers, but these were only partially successful.

The ability of *U. cinerea* to identify prey at some distance has contributed to its broad geographic distribution and biological success, primarily because of the extensive areas of bottom over which food is available to it by crawling. However, at what geographic distance the snail can identify a particular species of prey among others has not been reported. No one has followed the movements of these snails underwater in nature in response to distance attractants. Nor has the effect of gradients of salinity, temperature, light, and other ecological factors on the response and rate of movement of snails to specific prey been investigated. Another unanswered question is whether these snails are guided by a gradient of the attractant created by dilution in seawater, by attenuation of potency brought on by instability of the attractant, by interaction of the attractant with chemical and physical factors in the environment, or by a combination of these.

Our experiments demonstrated that minute quantities of a chemical cue in the fluid from the mantle cavity of oysters stimulate *U. cinerea* to mount oysters and to initiate penetration. Whether this response is reinforced by the same chemical (or some factor associated with it), which may be adsorbed to the shell, has not been demonstrated. Valvular movements of prey probably also reinforce the close-range chemical signal but are not necessary to initiate penetration. Thus closed oysters from which fluid escapes from the mantle cavity are vulnerable to attack, and under these conditions closure is no insurance against predation. We did not simulate the normal opening–closing rhythm of oysters with empty valves, but this would be worth doing to determine its possible role in the initiation of boring.

Valvular motion inhibits boring near and between the edges of the valves, but as soon as valves are clamped shut artificially, snails penetrate at the edges, the frequency of excavation seeming to increase with the rate of seepage of attractant from the mantle cavity of the prey. This response to valvular activity evolved, no doubt, from the consequence of pinching the proboscis between active valves. The behavior undoubtedly has survival value. Oysters irritated by insertion of the proboscis shut the valves tightly, clamp the proboscis between them, and remain closed as long as the irritation persists. Thus immobilized, the snail may lose its proboscis by tearing away or may die in place. Proboscides lost by clean experimental amputation regenerate fully in a short time (Carriker, 1961); thus a snail with its proboscis cleanly pinched off would recover. To what extent proboscides regenerate which are roughly torn off or rooted out of the cephalic hemocoel is not known.

U. cinerea can complete penetration of a valve in the absence of a live oyster, provided boring is triggered by attractant from an active live prey. The snail continues to bore the empty shell without further stimulation from the live oyster. Our discovery resulted in the development of the valve model, which provides a means for viewing the behavior of penetration of the shell at breakthrough and is the only technique available for collection of ABO secretion from normally functioning snails.

Shall penetration is accomplished by an intricate chemical–mechanical process in which the ABO secretes a substance which dissolves shell at the site of penetration and a minor portion of the weakened partly dissolved shell is removed by the radula and swallowed. The extent to which shell is utilized by these snails in calcium metabolism has not been explored; this could be determined by the use of isotopic calcium. Alternation of successive short periods of rasping and long periods of chemical activity results in the excavation of the borehole. Marginal teeth of the radula are about twice as hard as the rachidian teeth, and the latter are about the same hardness as oyster shell. Analysis of the teeth showed that the average elemental composition of both rachidian and marginal teeth is similar, that calcium is the major constituent, and that strontium and silicon are present as major to trace constituents. The distribution of chemical compounds in the teeth relative to hardness has not been studied.

Excavation of shell involves the close interaction of the proboscis, propodium, and ABO, in this order, in a predictable cycle which repeats itself continuously throughout the process of boring of each borehole. The ABO provides the shell-dissolving solution, the proboscis removes weakened shell, and the propodium, in addition to providing sensory information and breaking off thin edges of the borehole at breakthrough, voids seawater from the incomplete borehole prior to entrance of the ABO, thus minimizing dilution of the secretion.

While in position in the borehole, the ABO secretes sufficient solubilizing substance to dissolve a thin layer of shell at the bottom. In most cases this is adequate to obliterate the rasp marks of the previous rasping period. Preliminary characterization of the secretion has been reported (Carriker *et al.*, 1969). As penetration proceeds and the bottom of the borehole approaches the inner surface of the valve, there appears just before breakthrough a thin, often nearly transparent window of partially disolved shell through which the activities of boring are visible. The pattern of rasping is variable and generally asymmetric. The ABO, after conspicuous initial movements as it is positioned in the borehole, continues to pulsate gently. Initial breakthrough may be made by chemical dissolution or by rasping, and by some snails the radula is used to break off the thin edges of the emerging hole.

Study of the duration and frequency of rasping periods of the radula in oyster shell by means of amplified sounds obtained through a contact microhydrophone disclosed that after the start of penetration and up to the time of breakthrough the rasping periods were surprisingly uniform in duration, averaging abut 50 sec each. Following breakthrough, the duration of rasping periods became widely erratic as the snails rasped at the thin edges of the shelf of the hole. Variation of the interval of ABO activity averaged roughly 20 min at the start of penetration and lengthened to about

30 as boring progressed. The rate of shell penetration was about 0.32 mm per day. Yet to be determined is the rate of penetration of the shell of the oyster by representative sizes of *U. cinerea* under a variety of ecological conditions including a range of salinities, temperatures, and exposures to air simulating intertidal conditions. From a comparative behavioral point of view it would also be instructive to extend the study to include other species of shell-boring gastropods and several species of prey.

Borehole morphology is the product mainly of dissolution by the ABO secretion; thus the form and size of the borehole generally mirror the shape and size of the secretory disc of the ABO. The radula serves no part, or possibly only a minor one, in shaping the borehole. The radula functions in removing shell fragments and debris, particularly at the beginning and termination of excavation, and in increasing the surface area of the bottom of the borehole prior to each application of the ABO. The extent to which rasping actually accelerates the rate of shell penetration is unknown. Radular activity may also serve a tactile function, giving the snail information on the increasing depth of the borehole and its final breakthrough. To date, study of the form of the borehole has been undertaken only in prey shell. Shell, however, varies in composition and architecture and tends to influence the form of the borehole. This difficulty could be circumvented by study of the variation of boreholes made by individual snails in an artificial homogeneous calcareous material coated on the valves of live oysters.

Within the range of shell heights of 14–42 mm in both sexes of *U. cinerea*, increase in height is accompanied by a roughly proportionate increase in radular width and diameter of borehole. The diameter of boreholes, however, varies measurably among snails of a given size and of a given radular width, indicating that a range of sizes of ABOs exists in snails of the same size group. This is additional evidence for the negligible part played by the radula in the shape and size of the borehole. The size of boreholes excavated by the same snail during a short period likewise varies. Increases do not necessarily follow periods of feeding, nor decreases periods of fasting. The reasons for this are not clear and may relate to variations in the blood pressure of the snail. Deepest incomplete boreholes by large snails range in depth from 4.3 to 5.4 mm, suggesting that these depths may approach maximum penetrability by this species. From this, one would suppose that old thick-shelled oysters would be relatively safe from predation. This is not necessarily the case, as some snails may bore near the edge of the valves where the shell is thin enough to allow complete penetration. Of greater protection to large oysters than thick shell is the proximity of young oysters, which are more attractive than old oysters and thus serve as biological buffers to predation.

The capacity of *U. cinerea* to detect and locate fresh oyster flesh, though not as keen as for attractant from live prey, is highly developed. Feeding is carried out jointly by the rachidian and marginal teeth, which form a series of transverse five-cusped grasping structures. These tear loose morsels of flesh and convey them into the buccal cavity, where suction from the esophageal bulb sucks them into the esophagus.

There are several reports in the literature on the rate of feeding by *U. cinerea* based on the number of young oysters bored and consumed (Carriker, 1955*a*). No one, however, has yet devised a method for determining the actual rate of ingestion of food exclusive of boring time. Federighi (1931) suggested that this snail releases a toxic substance in the final stages of boring and initial feeding which kills the oyster. This has not been confirmed. Why these snails should employ a lethal toxin in the penetration of such docile prey is not clear.

Even in its long evolutionary history, so far as we know, the oyster has evolved no effective defense against approach and penetration by *U. cinerea.* Closing its valves brings the oyster only temporary respite, as in time it will open to feed and in so doing will release metabolites which will attract snails to it. Galtsoff's (1964) observation that addition of extract of the flesh of *U. cinerea* to the mantle edge of the oyster caused the tentacles to react strongly suggests that the oyster may recognize the metabolites of approaching snails as "danger signals." Such recognition, if true, although not seeming to serve the immobile sedentary oyster any useful purpose, may have resulted anyway from the long evolutionary association of prey and predator.

The effectiveness of the shell-penetrating mechanism has been amply demonstrated by the costly depredations by *U. cinerea* of oysters on commercial beds. The mechanism allows *U. cinerea* to bore and feed on otherwise generally inaccessible prey often much larger than themselves protected by their own shells and at least for a time by the valves of the prey. After an oyster gapes as the result of the first boring, other snails in the vicinity join in the meal, moving in between the valves without the need of boring. Thus food in a large oyster, which would have decayed before the single snail could have consumed it, is eaten by other snails and the entire local population of predators benefits. Only when prey valves gape widely and small crabs and fish join the meal is there danger of loss of the proboscis to the newcomers. Such proboscisectomies, however, do not seem to occur frequently.

All species of Muricidae which have been examined to date possess the typical proboscis and ABO described for *U. cinerea* (Carriker, 1961). The behavior of boring by *Eupleura caudata* and *E. c. etterae* has also been examined and is similar to that of *U. cinerea.* Whether the boring behavior is similar for all other muricid gastropods can only be confirmed after many more species have been studied, but morphology would suggest that it probably is. The morphological mechanism of shell penetration of the naticid boring gastropods differs from that of the Muricidae primarily in that the

ABO is located under the ventral tip of the proboscis rather than in the foot; scattered information on the Naticidae tentatively suggests that their behavior of penetration approximates that of the Muricidae (Carriker and Yochelson, 1968). The ubiquity of gastropod boreholes in Recent and fossil mollusk shells is an index of the abundance, world-wide distribution, and success of gastropod boring species. Knowledge of the behavior of hole boring by other invertebrates and by the lower plants is still too scanty to allow comparison with that of the gastropods (Carriker *et al.*, 1969).

The boring habit in gastropods probably arose first in the Naticidae in Upper Cretaceous times. The Jurassic and Lower Cretaceous naticaceans must have fed in some other manner. Muricids first appeared in the Lower Cretaceous, but borings attributable to muricids seem first to have occurred in the Upper Cretaceous (Sohl, 1969). Earlier borings superficially resembling gastropod borings in Paleozoic and pre-Upper Cretaceous brachiopods and a few mollusks are of unknown origin (Carriker and Yochelson, 1968; Sohl, 1969). Thus the habit of shell boring by snails is very old, originating probably between 60 million and 130 million years ago. How long before the Cretaceous the behavior was evolving is difficult to know. It is probable that shell-boring snails arose from nonboring carnivorous gastropods.

Penetration of shell of prey by *U. cinerea* is of fundamental biological, medical, and dental interest because it involves the chemical dissolution of both the mineral crystals and organic matrix of shell. Dissolution of mineralized tissues, including skeletons, teeth, and other specialized calcified structures is performed by many species throughout the animal kingdom, including man and the lower plants (Carriker *et al.*, 1969; Sognnaes, 1963). Knowledge gained in the study of a basic process in one species is likely to contribute to similar investigations of other species, including man.

ACKNOWLEDGMENTS

John W. Blake and Langley Wood assisted in the investigations at the Institute of Fisheries Research, and Eva S. Montiero and Mark Sherman aided in those at the Marine Biological Laboratory. Barry Martin developed the contact microhydrophone and amplifier for recordings the rasping sounds of boring snails and assisted in their utilization. James J. Bussey, Jarrell Ash Division, Fisher Scientific Company, conducted the chemical analysis of radular cusps. Michael Castagna, Virginia Institute of Marine Science, supplied live *U. c. follyensis* from the eastern shore of Virginia. R. A. Boolootian loaned a moving-picture camera, and E. R. Baylor provided a tape recorder. Photographs in Figs. 3, 6, 7, and 37 were taken by P. J. Oldham. Scanning micrographs in Figs. 11 and 19 were taken by F. R. Round.

John W. Blake, Paul S. Galtsoff, Bori L. Ollla, David M. Pratt, Howard E. Winn, and Langley Wood kindly reviewed the manuscript and offered many valuable suggestions.

The research in North Carolina was supported in part by a grant from the U.S. Fish and Wildlife Service and that at the Marine Biological Laboratory by Public Health Service Research Grant DE 01870 from the National Institute of Dental Research.

Acknowledgment is gratefully made for the many courtesies and generous assistance which made this study possible. Systematics–Ecology Program Contribution No. 229.

REFERENCES

Alexander, C. G., 1970, The osphradium of *Conus flavidus*, *Mar. Biol.* **6**: 236–240.

Baker, B. B., 1951, Interesting shells from the Delmarva Peninsula, *Nautilus* **64**: 73–77.

Blake, J. W., 1960, Oxygen consumption of bivalve prey and their attractiveness to the gastropod, *Urosalpinx cinerea*, *Limnol. Oceanograph.* **5**: 273–280.

Blake, J. W., 1961, Preliminary characterization of oyster metabolites attractive to the predatory gastropod, *Urosalpinx cinerea*, Ph. D. thesis, University of North Carolina, Chapel Hill, 46 pp.

Blake, J. W., 1962, Preliminary characterization of oyster metabolites attractive to the predatory gastropod, *Urosalpinx cinerea*, *Dissertation Abst.* **23** (2).

Blake, J. W., 1966, Inherent differences between populations of the oyster-drilling gastropod, *Urosalpinx cinerea*, reared under laboratory conditions, 2nd Internat. Oceanogr. Congr., Abst. Papers (No. 45, SIIc), Moscow.

Carriker, M. R., 1943, On the structure and function of the proboscis in the common oyster drill, *Urosalpinx cinerea* Say, *J. Morphol.* **73**: 441–506.

Carriker, M. R., 1955*a*, "Critical Review of Biology and Control of Oyster Drills *Urosalpinx* and *Eupleura*", *U.S. Fish Wildlife Serv. Spec. Sci. Rep.—Fish.* **148**: 1–150.

Carriker, M. R., 1955*b*, Seasonal vertical movements of oyster drills (*Urosalpinx cinerea*), *Proc. Nat. Shellfish. Assoc.* **45**: 190–198.

Carriker, M. R., 1957, Preliminary study of behavior of newly hatched oyster drills, *Urosalpinx cinerea* (Say), *J. Elisha Mitchell Sci. Soc.* **73**: 328–351.

Carriker, M. R., 1961, Comparative functional morphology of boring mechanisms in gastropods, *Am. Zoologist* **1**: 263–266.

Carriker, M. R., 1969, Excavation of boreholes by the gastropod, *Urosalpinx:* An analysis by light and scanning electron microscopy, *Am. Zoologist* **9**: 917–933.

Carriker, M. R., and Martin, B., 1965, Analysis of shell boring behavior of muricid gastropod, *Urosalpinx cinerea* (Say), by means of color motion picture and microhydrophone recording of radular sounds, *Am. Zoologist* **5**: 645.

Carriker, M. R., and Smith, E. H., 1969, Comparative calcibiocavitology: Summary and conclusions, *Am. Zoologist* **9**: 1011–1020.

Carriker, M. R., and Van Zandt, D., 1964, Use of polished mollusk shell for testing demineralization activity of accessory boring organ of muricid boring gastropods, *Biol. Bull.* **127**: 365.

Carriker, M. R., and Yochelson, E. L., 1968, Recent gastropod boreholes and Ordovician cylindrical borings, Contrib. Paleontol., Geol. Surv. Prof. Paper 593–B, B1–B26.

Carriker, M. R., Charlton, G., and Van Zandt, D. 1967, Gastropod *Urosalpinx:* pH of accessory boring organ while boring, *Science* **158**: 920–922.

Carriker, M. R., Scott, D. B., and Martin, G. N., 1963, Demineralization mechanism of boring gastropods, *in* "Mechanisms of Hard Tissue Destruction" (R. F. Sognnaes, ed.) Publ. No. 75, pp. 55–89, Am. Assoc. Advan. Sci., Washington, D. C.

Carriker, M. R., Smith, E. H., and Wilce, R. T., 1969, Penetration of calcium carbonate substrates by lower plants and invertebrates, *Am. Zoologist* **9(3)**: 629–1020.

Chetail, M., and Fournié, J., 1970, Mécanisme de perforation chez *Thais lapillus* L. (Gastéropode, Prosobranche, Muricidé): Mise en évidence d'une entrée d'ions calcium durant l'activité de l'organe de perforation, *Compt. Rend. Acad. Sci. Paris* **271**: 118–121.

Day, J. A., 1969, Feeding of the cymatiid gastropod, *Argobuccinum argus*, in relation to the structure of the proboscis and secretions of the proboscis gland, *Am. Zoologist* **9**: 909–916.

Federighi, H., 1931, Studies on the oyster drill (*Urosalpinx cinerea* Say), *U.S. Bur. Fish. Bull.* **47**: 83–115.

Fretter, V., and Graham, A., 1962, "British Prosobranch Molluscs; Their Functional Anatomy and Ecology," Royal Society, London 755 pp.

Galtsoff, P. S., 1964, The American Oyster *Crassostrea virginica* Gmelin, Fish. Bull., Fish Wildlife Serv,. No. 64, 480 pp.

Galtsoff, P. S., Prytherch, H. F., and Engle, J. B., 1937, Natural history and methods of controlling the common oyster drills (*Urosalpinx cinera* Say and *Eupleura caudata* Say), U.S. Bur. Fish. Cir. No. 25, pp. 1–24.

Hancock, D. A., 1959, The biology and control of the American whelk tingle *Urosalpinx cinerea* (Say) on English oyster beds, Great Britain Ministry Agriculture, Fisheries and Food, Fishery Invest., Ser. 2, Vol. 22, No. 10, 66 pp.

Hanks, J. E., 1957, The rate of feeding of the common oyster drill, *Urosalpinx cinerea* (Say), at controlled water temperatures, *Biol. Bull.* **112**: 330–335.

Haskin, H. H., 1940, The role of chemotropism in food selection by the oyster drill, *Urosalpinx cinera* Say, *Anat. Rec.* **78**: 95.

Haskin, H. H., 1950, The selection of food by the common oyster drill, *Urosalpinx cinerea* Say, *Proc. Nat. Shellfish. Assoc.* **1950**: 62–68.

Kohn, A. J., 1961, Chemoreception in gastropod molluscs, *Am. Zoologist* **1**: 291–308.

Korringa, P., 1951, The shell of *Ostrea edulis* as a habitat, *Arch. Neerl. Zool.* **10**: 31–152.

Kumpf, H. E., 1964, Use of underwater television in bio-acoustic research, *in* "Marine Bio-Acoustics," pp. 45–47, Pergamon Press.

Manzi, J. J., 1970, Combined effects of salinity and temperature on the feeding, reproductive, and survival rates of *Eupleura caudata* (Say) and *Urosalpinx cinerea* (Say) (Prosobranchia: Muricidae), *Biol. Bull.* **138**: 35–46.

Myers, T. D., 1965, A comparative study of size variation in the Atlantic oyster drill, *Urosalpinx cinerea* (Say), M. A. thesis, University of North Carolina, Chapel Hill, 77 pp.

Nylen, M. U., Provenza, D. V., and Carriker, M. R., 1969, Fine structure of the accessory boring organ of the gastropod, *Urosalpinx*, *Am. Zoologist* **9**: 935–965.

Orton, J. H., 1930, On the oyster drill in the Essex estuaries, *Essex Nat.* **22**: 298–306.

Person, P., Smarsh, A., Lipson, S. J., and Carriker, M. R., 1967, Enzymes of the accessory boring organ of the muricid gastropod, *Urosalpinx cinerea follyensis*. I. Aerobic and related oxidative systems, *Biol. Bull.* **133**: 401–410.

Pope, T. E. B., 1910–1911, The oyster drill and other predatory mollusca, Unpublished Rep. U.S. Bur. Fish., Washington, D.C., 47 pp.

Runham, N. W., 1961, The histochemistry of the radula of *Patella vulgata*, *Quart. J. Microscop. Sci.* **102**: 371–380.

Runham, N. W., 1963, A study of the replacement mechanism of the pulmonate radula, *Quart. J. Microscop. Sci.* **104**: 271–277.

Sizer, I. W., 1936, Observations on the oyster drill with special reference to its movement and to the permeability of its egg case membrane, Unpublished Rep. U.S. Bur. Fish., Washington, D.C.

Smarsh, A., Chauncey, H. H., Carriker, M. R., and Person, P., 1969, Carbonic anhydrase in the accessory boring organ of the gastropod, *Urosalpinx*, *Am. Zoologist* **9**: 967–982.

Sognnaes, R. F., 1963, Mechanisms of hard tissue destruction, Am. Assoc. Adv. Sci., Publ. No. 75, 764 pp.

Sohl, N. F., 1969, The fossil record of shell boring by snails, *Am. Zoologist* **9**: 725–734.

Stauber, L. A., 1943, Ecological studies on the oyster drill, *Urosalpinx cinerea* in Delaware Bay, with notes on the associated drill, *Eupleura caudata*, and with practical consideration of control methods, Unpubl. Rep., Oyster Research Lab., N.J.

Webb, K. L., and Wood, L., 1967, Improved techniques for analysis of free amino acids in seawater, *in* "Automation in Analytical Chemistry, Technicon Symposia," pp. 440–444, Mediad, 1967, New York.

Wood, L. H., 1965, Physiological and ecological aspects of prey selection by the marine gastropod, *Urosalpinx cinerea* (Say), Ph. D. thesis, Cornell University, 216 pp.

Wood, L. H., 1966, Determination of free amino acids in seawater, *in* "Automation in Analytical Chemistry, Technicon Symposia," pp. 652–655, Mediad, 1966, New York.

Wood, L. H., 1968, Physiological and ecological aspects of prey selection by the marine gastropod, *Urosalpinx cinerea* (Prosobranchia: Muricidae), *Malacologia* **6**: 267–320.

Young, D. K., 1969, *Okadaia elegans*, a tube-boring nudibranch mollusc from the central and west Pacific, *Am. Zoologist* **9**: 903–907.

INDEX

This is a joint index for Volumes 1 and 2. Pages 1–244 will be found in Volume 1 and pages 245–492 will be found in Volume 2.

Abudefduf, 298–299
Acanthocybium solandri, 260
Acanthuridae, 278
Activity (*see also* Photoperiod, Rhythms)
  adaptive, in shore-living arthropods, 2–12
  daily, in *Uca,* 61, 63
  daily changes, in *Pomatomus,* 307–311, 324
  in fishes, 304
  nocturnal
    in *Ocypode,* 61, 63
    in *Uca,* 63
  seasonal changes
    in *Gadus,* 305
    in *Pomatomus,* 311–315, 324, 325
  tidal periodicity, in *Uca,* 61, 63
Adaptive mechanisms in shore-living arthropods, 2, 3, 7, 8, 10
Aggression (*see also* Ritualization)
  in crustaceans, 99, 101, 104–115
  in *Eupomacentrus,* 300, 442, 446, 447, 449, 453, 464–465
  in hermit crabs, 116–117
  in *Opsanus,* 361
  shell-fighting, in hermit crabs, 101, 106
  territorial-aggressive signals, 100, 119
Agonistic patterns (*see* Aggression)
Alpheus, 102
Amphipods, 1–12, 25, 39, 41–44, 64, 100
Anchoviella, 438
Ants, 9, 127, 153
Apogonidae, 278
Arachnids, 1, 3, 8–9, 11–12, 18, 25, 40, 42, 43
Arctosa, 25, 43
  A. cinerea, 8
  A. perita, 8
  A. variana, 8–9, 40
Arthropods, 1–55, 64, 100
Association, with drifting objects
  in *Euthynnus,* 278
  in tunas, 246, 258
Atlantic mackerel, 272
Auxis thazard, 262, 266

Bagre marinus, 377
Bagrids, 376
Balanus, 161
Barents sea cod, 305
Barbet, 428
Barnacles, 159, 160
Barracudas, 297, 438
Barth's organ, 69

Bathygobius soporator, 377, 379, 424–425, 438–439
Bees, 9, 127, 136, 153
Beetles, 1, 9
Beluga, 437
Bicolor damselfish, 278–300, 381, 439–455
Bigeye tuna, 269–271
Biological clocks (*see* Rhythms)
Birgus, 10
Black skipjack, 258
Blue-backed manakin bird, 428
Blue mussel, 159, 160, 183
Blue runner, 439
Bluefish, 303–325
Bonito, 272
Boring behavior (*see* Shell-boring)
Bottlenose dolphin, 251, 436, 470
Brachyura, 25, 99–101
Bucephala clangula, 119–120
Bullfrog, 374, 425–426

Calcibiocavites, 157–243
Calcinus ornatus, 114
  C. tibicen, 104, 114
  C. verrilli, 114
California sea lion, 436, 470, 477–490
Calliactis parasitica, 115
Callinectes, 2
Cannibalism, in *Ocypode,* 64
Capture
  and maintenance of tunas, 258, 262–272
  "puffy snout," 271–272
Carabidae, 1, 9
Carangidae, 259, 267, 279, 297, 438, 439
Caranx, 438
  C. fusus, 439
Carcharhinus falciformis, 457, 461
  C. longimanus, 259
C. falciformis (*cont.*)
  C. obscurus, 457
  C. springeri, 457
Carcinus, 11
Cardinal, 426
Cardinalfishes, 278
Cardisoma, 10, 11
Catfishes, 376–377
Cetaceans, 251, 436–439, 469–470, 490
Chemoreception
  in crabs, 102
  in *Urosalpinx,* 160–243
Chromatic patterns (*see* Color patterns)
Chromis, 280, 298–299
Cleaning symbiosis
  in *Periclimenes,* 100–101
  in tunas, 257–258
Clibanarius anomalus, 114
  C. antillensis, 114
  C. crythropus, 114
  C. cubensis, 114
  C. panamensis, 114
  C. tricolor, 114
  C. vittatus, 104–105, 107–114, 116, 120
Coenobita, 10, 101
Coenobitidae, 1
Color patterns
  in *Coryphaena,* 250
  in *Eupomacentrus,* 442, 445, 447, 449, 450, 451, 452, 453, 464
  in *Katsuwonus,* 248, 250, 264
  in tunas, 267, 273
Common dolphin, 437
Common sole, 304
Communication (*see also* Ritualization, Sound, Sound detection, Sound production)
  analysis of,
    descriptions of displays, 64–67,

Communication (*cont.*)
126, 127, 149
information theory, 126
quantitative, 149–154
information transfer, 373, 379
language level, 381
in marine crustaceans, 61–64, 99–102
nonritualized, 112, 114–115
ritualized, 78–79, 99–100, 103–108, 118
social behavior, in *Uca,* 61–63
symbolic level of, 381
tactile stimuli, in *Uca,* 101–102
Coryphaena hippurus, 250, 260, 261, 262, 267
Cottus scorpius, 374
Courtship
in *Eupomacentrus,* 442, 443, 444, 445, 446, 447, 449, 453, 463–464
in *Euthynnus,* 258
in *Katsuwonus,* 257
in *Ocypode,* 64
in *Opsanus,* 374, 377, 379, 380
in *Sarda,* 272
in tunas, 273
in *Uca,* 62, 63, 78–79, 100, 195, 462
Crassostrea virginica, 158–243
Cryptomonas, 163
Current
effect on feeding
in *Eupomacentrus,* 285–286, 290–296
in reef fishes, 299
effect on distance to habitat
in *Eupomacentrus,* 283–290, 293, 295–296

Damselfishes, 278–300, 373, 381, 439–455
Dardaninae, 114
Dardanus arrosor, 114
D. callidus, 114
D. venosus, 114
Dascyllus albisella, 280, 298
Decapods, 1, 3, 10–38, 40–44, 60–94, 99–105, 127, 140–155, 462, 463
Delphinapterus leucas, 437
Delphinus delphis, 437
Diogenes pugilator, 114
Diogenidae, 114
Displays (*see* Aggression; Communication; Ritualization)
Dolphin (cetacean) (*see also* Porpoise), 251, 436, 437, 439, 470
Dolphin (fish) 250, 260, 261, 262, 267
Dotilla, 10, 100
Drums, 2, 278–279, 376, 379
Dusky shark, 457
Dyschirus numidicus, 9

Echolocation, by cetaceans, 482
Echosounders, to track tunas, 252
Elagatis bipinnulata, 267
Elasmobranchs, 259, 373, 380, 456–461
Energy flow, in reef ecosystems, 278–280, 206–299
Environmental features, tuna behavior affected by, 246
Enzymes, used in shell-boring, by *Urosalpinx,* 193
Escape responses, in *Uca,* 12–25
Eschrichtius glaucus, 437
Eumetopias jubata, 470–476, 487, 489
Eupomacentrus partitus, 278–300, 381, 439–455
Euthynnus affinis, 252, 258, 262, 264, 267, 270, 272
E. lineatus, 258

Evolution, of boring molluscs (*Urosalpinx*), 240–241
Eye (*see also* Vision)
  morphology and histology, of tuna, 271
  movement of cones in *Pomatomus*, 305, 323, 324
  of pinnipeds, 482, 488-489
  retinal adaption in fishes, 305

Fantail warbler, 428
Feeding
  current effect on
    in reef fishes, 299
    in *Eupomacentrus*, 285–286, 290–296
  in *Eupomacentrus*, 285–286, 293, 295–296, 442
  light effect on, in Pomacentridae, 281, 285–286, 290–296
  in *Petrochirus*, 105
  in pinnipeds, 469, 481
  in *Pomatomus*, 307, 324
  in reef fishes, 278–280, 297–299
  in tunas, 246, 248–250, 264, 267, 272, 273
  turbidity effect on
    in *Eupomacentrus*, 297
    in *Phoca*, 489
  in *Uca*, 62
  in *Urosalpinx*, 230–235
Fiddler crabs, 2, 3, 10–38, 40–44, 60–79, 82, 85, 91–94, 100–102, 462
Finch, 426
Fins, hydrodynamics of pectoral, in tunas, 273
Fishing gear, responses to, by *Katsuwonus*, 251
Freckled driftfish, 261
Frigate mackerel, 262–266
Frillfin goby, 377, 379, 424–425, 438, 439

Gadus morhua, 305
Galeichthys felis, 377
Gasterosteus aculeatus, 304, 305
Gastropods, 157–243
Gecarcinus, 10
Genotypic variation, in *Urosalpinx*, 159
Ghost crab, 1, 3, 10–12, 25, 38, 41, 60, 61, 63–64, 67–69, 71, 100
Ginglymostoma cirratum, 457
Goatfishes, 278–279
Goldeneye duck, 119–120
Goldfish, 380
Goniopsis cruentata, 10, 25, 41
Gonodactylus, 140–155
  G. bredini, 103, 127, 140–155
  G. spinulosus, 127, 140–155
Grampus, 439
Grapsidae, 1, 10, 11, 25, 41, 64, 100
Grapsus, 10
Gray snapper, 298
Gray whale, 437
Groupers, 279, 376
Grouping
  as a pooling of directional information, 18
  in *Uca*, 18, 29
Grunts, 279, 439, 460
Gryllotalpidae, 1

Habitat (*see also* Shelter-seeking), distance to
  current and light effects on, in *Eupomacentrus*, 280, 283–290, 293, 295–296
  feeding and, in *Eupomacentrus*, 285–286, 293, 295–296
  in Pomacentridae, 280–281
Haemulon album, 439, 460
Halichoeres bivittatus, 439
Harbor seal, 436, 470–477, 487, 489–490
Hearing (*see* Sound detection)

Hibernation, in *Urosalpinx*, 159
Heloecious, 100
  H. cordiformis, 102
Hemigrapsus, 10
  H. oregonesis, 11
Hemiplax, 100
  H. hirtipes, 99
  H. latifrons, 99
Hermit crab, 25, 101–105, 127, 141, 150–152
Holocentridae, 377, 379, 380, 381, 453
Holocentrus coruscus, 377, 379
  H. rufus, 377, 381, 453
Homarus, 101

Indigo bunting, 426
Information theory
  in communication analysis, 127, 136–138, 150–154
  Dancoff's principle, 130, 137
  limitations of, in behavior analysis, 139–140
  measures of communication, 130, 131, 132
  methods relative to behavior, 135–136, 140–148
  Shannon–Wiener formulation, 129
  symbols and terms, 127, 128, 130, 133, 134
Insects, 9, 11
Isopods, 2–3, 7–9, 11, 25, 39, 42, 43

Jacks, 259, 267, 279, 297, 438, 439

Katsuwonus pelamis, 248–252, 255, 257–258, 262–264, 266–269, 271–273
Katydids, 150, 429, 430
Kawakawa, 252, 258, 262, 264, 267, 270, 272
Killer whale, 437, 469, 490

Laboratory facilities
  aquaria, 306
  tanks and ponds, for catching and maintaining tunas, 262–274
Labridae, 278, 439
Lagenorhynchus obliquidens, 436, 469, 490
Laniarius, 427, 428, 432
  L. aethipicus major, 428
Lateral line
  of *Katsuwonus*, 271
  of *Opsanus*, 374
Lemon shark, 458
Leptonychotes weddelli, 437
Light (*see also* Photoperiod, Orientation)
  activity affected by, in fishes, 304, 324
  distance to habitat affected by, in *Eupomacentrus*, 283–290, 293, 295–296
  feeding affected by
    in *Eupomacentrus*, 285–286, 290–296
    in shore fishes, 279
  phasic component, 316–318
  polarized light perception, in arthropods, 11
  simulated diurnal changes, 306, 307, 315–318, 320–322
  visual acuity affected by,
    in pinnipeds, 481–482
    in *Zalophus*, 282–288
Lobsters, 25, 99, 101, 463
Locomotory movements
  in *Clibanarius*, 109–114
  in *Eupomacentrus*, 442, 445, 446, 447, 448, 449, 452, 453, 454, 455, 463–465
  in *Petrochirus*, 109–114
  in *Uca*, 36–37, 60–64
Lutjanus analis, 439
  L. griseus, 298

Lycosid spiders, 1, 3, 8–9, 11–12, 18, 42, 43

Mantis shrimp, 103, 127, 140–155
Margate, 439, 460
Mesogastropoda, 161
Mictyridae, 1, 10, 99
Mictyris, 10
  M. longicarpus, 99
Migration (*see also* Rhythms)
  in fish, 304, 305
  photoperiod effect, in *Pomatomus,* 324, 325
  in shore-living isopods, 7
Mockingbird, 427
Models
  to elicit display, 98–99, 102, 108
  of prey, 178–185
Mole crickets, 1
Movement (*see* Aggression, Locomotory movements, Ritualization)
Mullidae, 278–279
Muricidae, 157–243
Mutton snapper, 439
Mya arenaria. 200, 205
Mycteroperca bonaci, 377
Mynah bird, 427
Myripristis jacobus, 376–377
Mytilus edulis, 159, 160, 183

Naucrates ductor, 259
Negaprion, 458
Neogastropoda, 161
Nesting patterns, in *Eupomacentrus,* 442, 463
Northern fur seal, 489
Notropis analostanus, 377, 379, 425
  N. lutrensis, 425
  N. venustus, 425
Nucella lapillus, 205
Nurse shark, 457

Ocypode, 10, 38, 43, 60
  O. ceratophthalma, 11–12, 25, 41, 61, 63–64, 67–68
Ocypode (*cont.*)
  O. laevis, 61, 67–68
  O. quadrata, 61, 63, 67–69, 71
  O. saratan, 11, 41, 61, 64, 100
Ocypodidae, 1, 3, 10–12, 25, 38, 41, 60, 61, 63–64, 67–69, 71, 100
Ocyurus chrysurus, 297, 439, 460
Olfaction, in tunas, 258, 262, 271, 273
Olfactory organ, in *Katsuwonus,* 271
Omophron limbatum, 9
Opsanus tau, 361–376, 379–383, 386–431, 439
Orcinus orca, 437, 469, 490
Orchestia gammarella, 5
  O. mediterranea, 4, 6, 39
Orchestoidea benedicti, 4, 39
  O. corniculata, 4–5, 39
Orientation
  in pinnipeds, to target, 470, 476–477, 479
  in shore-living arthropods, 1–44
    celestial, 7–10, 13, 15, 18, 27, 29, 31–33, 37, 38, 42
    detour integration, by *Uca,* 14
    directional preference, in decapods, 12, 24
    during ontogeny, in *Uca,* 25
    innate, in amphipods, 6, 24
    kinesthesis, in decapods, 11, 13, 15
    landmark, 5, 8, 11, 13, 15, 16, 18, 24, 27, 29, 43
    lunar, 5
    photo-orientation, in *Uca,* 31, 35, 43
    polarized light, 8, 11, 12, 15, 18, 24, 35, 37
      sun compass, 8, 34, 42, 44
  Ostariophysids, 380
  Oxygen consumption, in muscles of tuna, 271
  Oysters, 158–243

Pachygrapsus, 10, 102
Pacific bonito, 272
Pacific mackerel, 272
Paederus rubrothoracicus, 9
Paguridae, 25, 101–105, 127, 141, 150–152
Paguristes anomalus, 115
  P. cadenali, 114
  P. depressus, 114
  P. grayi, 114
  P. hummi, 115
  P. morrei, 114
  P. oculatus, 114
  P. punticeps, 107, 114
  P. sayi, 114
  P. spinnipes, 115
  P. triangulatus, 115
  P. tortugae, 114
Pagurus bernhardus, 102
  P. longicarpus, 25
Panulirus, 101
  P. argus, 25, 463
  P. interruptus, 463
Parrot, 427
Parrotfishes, 278, 304
Patella vulgata, 193
Periclimenes, 100
Petrochirus diogenes, 104–116
Phaleria provincialis, 9
Phoca vitulina, 436, 470–477, 487, 489–490
Pholidoptera girseoaptera, 431, 432
Photographic analysis, 109–115, 246, 248, 250–251, 258, 267
Photoperiod (*see also* Activity, Light, Rhythms)
  daily activity affected by, in *Uca,* 63
  migration affected by, in fishes, 305, 324, 325
  reproduction affected by, in *Gasterosteus,* 305
Photoperiod (*cont.*)
  seasonal activity affected by, in *Pomatomus,* 311–315, 324–325
Phoxinus laevis, 379
Pilotfish, 259
Pinnipeds, 436, 437, 460–491
Piranhas, 376–377
Plaice, 304
Pleuronectes platessa, 304
Podophthalmus, 11
Pomacentridae, 278–300, 373, 381, 439–455
Pomacentrus partitus (*see Eupomacentrus partitus*)
Pomadasyidae, 279, 439, 460
Pomatomus saltatrix, 303–325
Porichthys notatus, 376
Porpoise [*see also* Dolphin (cetacean)], 427
Portunas sanguinolentus, 102
Portunid crabs, 2, 15
Predation
  by Carangidae, 297
  on *Crassostrea,* 157–243
  on *Eupomacentrus,* 298
  on Pomacentridae, 297
  on shore-living arthropods, 1–3
  by Sphraenidae, 297
  by *Urosalpinx,* 157–243
Prionotids, 373, 376, 424, 438
Prionotus, 373, 424, 438
Psenes cyanophrys, 261
Proprioception, in decapods, 11
Pterophylla camellifolia, 150, 429, 430

Quail, 427, 428, 432

Rainbow runner, 267
Rana catesbeiana, 374, 425–426
Reef ecosystems
  energy cycling in, 278–280, 298

Reef ecosystems (*cont.*)
energy sources in, 278–279, 296–299
Reef shark, 457
Respiration
oxygen tolerance, in tunas, 271
Reproductive activity
in *Eupomacentrus,* 281, 296, 300
in *Pomacentridae,* 280–281
in *Opsanus,* 361, 366–367, 370–371, 382
in *Uca,* 61–63, 78
Rhesus monkey, 103, 127, 153
Rhincodon typus, 259
Rhizoprionodon, 457
Ritualization (*see also* Aggression, Communication, Sound production)
in *Clibanarius vittatus,* 104–105, 107–114, 116, 119–121
communication, 97–98, 118–122
definition, 97
in *Pagurus,* 115–118
in *Petrochirus,* 104–112, 114–115
in *Uca,* 61, 100–102, 119–121
waving display, in *Uca,* 61, 62, 63, 66, 78, 93, 95
Rhythms (*see also* Migration, Orientation)
amplitude, in *Pomatomus,* 312, 314
biological clock, in *Uca,* 19, 22, 24, 36, 38
daily
in *Pomatomus,* 303, 305, 308–311, 324
in *Solea,* 304
internal control, 303
in *Scarus,* 304
in *Pleuronectes,* 304
in *Pomatomus,* 305, 306, 317, 318, 321, 322, 324
in shore-living amphipods, 4–5
Rhythms (*cont.*)
light effect on, in *Pomatomus,* 315–324
seasonal, in *Pomatomus,* 303, 311–315, 324–325

Salinity
effect on activity, in *Urosalpinx,* 162
preference changes, in *Gasterosteus,* 304–305
Sarda chiliensis, 272
S. orientalis, 272
Satinfin shiner, 377
Scaridae, 278, 304
Scarites terricola, 9
Scarus coelestinus, 304
S. guacamaia, 304
Sciaenidae, 2, 278–279, 376, 379
Schooling
fish responses to sound, 438, 439
in *Pomatomus,* 303, 305, 307, 309–311, 313–316, 321–322, 324
in tunas, 246, 250, 255, 257, 276
Euthynnus, 267
Katsuwonus, 248–249
Scomber japonicus, 272
S. scombrus, 272
Scombridae, 245–274
Scorpaenidae, 279
Sea robin, 373, 424, 438
Seawater system (*see* Laboratory facilities)
Serranidae, 279, 376
Sesarma, 10
S. meinerti, 100
Sharpnose shark, 457
Shell-boring, in *Urosalpinx,* 157–243
boring organ, 185–243
Shelter-seeking (*see also* Habitat)
in shore-living arthropods, 2, 3, 13–21
Shrike, 427, 428, 432

Silky shark, 457, 461
Skeletonema, 163
Skipjack tuna, 248–252, 255, 257–258, 262–264, 266–269, 271–273
Slippery dick, 439
Snapping shrimp, 102
Social behavior (*see* Communication, Ritualization)
Solea vulgaris, 304
Sonar (*see also* Echolocation)
  to track tunas, 252, 254–255
Sound (*see also* Communication)
  biological significance of, 438, 442, 462
    in birds, 426–428, 431
    calling rate, 431
    in fishes, 418, 421–422, 424–425, 431
    in frogs, 425–426
    in individual recognition, 426, 427
    in insects, 429–431
    intensity, 431
    in *Opsanus,* 418, 421–422, 431
    in reproduction, 418, 421–422, 424–425, 426, 427
    in spacing of individuals, 429
    in species recognition, 425, 427
  communication
    in birds, 120, 426–429
    in fishes, 372, 373, 376, 379, 381, 424–426, 429–431
    in frogs, 425–426
    in insects, 429–431
    levels of interaction, 381
    in *Ocypode,* 64
    ontogeny of, in birds, 426
    in *Opsanus,* 386–387
    species discrimination, 373, 379
    territorial behavior, 377, 379, 380, 381, 382
    in *Uca,* 78–79
Sound detection (*see also* Communication)
  acoustico-lateralis system, in *Opsanus,* 372, 374
  and attraction response
    in elasmobranchs, 373, 456–461
    in *Opsanus,* 373, 374, 382, 383
    in teleosts, 438, 439, 460, 461
  in cetaceans, 436, 437
  control of behavior, of teleosts, 436, 438, 453
  directional orientation
    in elasmobranchs, 456
    in *Uca,* 12
  in elasmobranchs, 456–462
  in *Eupomacentrus,* 447
  frequency spectrum, of elasmobranchs, 456–462
  graded responses, 374, 375, 376, 377, 379, 381
  habituation
    in elasmobranchs, 456, 457, 459
    in *Eupomacentrus,* 452, 453
    in pinnipeds, 438
  localization, 373, 374, 380, 382, 383
  mechanically produced sound
    responses to
      in elasmobranchs, 456–461
      in invertebrates, 462
      in pinnipeds, 437–438
      in teleosts, 439
      in *Uca,* 69–77
    to track tunas, 252
  in *Ocypode,* 68, 69, 71, 94–95
  in *Opsanus,* 372, 374, 380, 383
  in pinnipeds, 436–438
  responses
    in *Eupomacentrus,* 442–455
    in invertebrates, 462–463
    in *Opsanus,* 421
    in *Prionotus,* 424
    in *Rana,* 426

Sound detection (*cont.*)
in teleosts, 438–455
in *Uca,* 78–92
thresholds, in *Opsanus,* 372, 374, 380, 381
in tunas, 262, 264, 273
Sound production (*see also* Communication)
ambulatory leg movements, in *Uca,* 64–66, 93
antiphony, 361, 422–423, 428–431
body thumping, in *Uca,* 66
calling rates, 120, 418–423, 429–431
in cetaceans, 436–437
countersinging, 428–429
duet calling, in birds, 428
in *Eupomacentrus,* 442–455
inhibition, in *Opsanus,* 423–424, 429–430
in invertebrates, 462–463
message content, 362, 375, 376, 377, 381, 382
mimicry, 419, 427–428
in *Ocypode,* 64, 67–68, 94–95
in *Opsanus,* 361–382, 386–426, 432, 439
in pinnipeds, 437–438, 470–471, 477, 479–480, 489
rapping, in crabs, 66, 67–68, 78–79, 85–92
rasping
in Ocypode, 67–68
in panulurid lobsters, 101
in *Urosalpinx,* 205, 209–210
signal systems
of *Opsanus,* 361, 375–379
of *Uca,* 61, 64–68, 78–79, 92–95
sound attenuation and speed, 96, 88, 380
swim bladder, 375, 377, 378, 381, 432
stridulation, 376, 377, 378, 379
Sound production (*cont.*)
in *Coenobita,* 101
in *Uca,* 66–67, 68
synchronous calling, in birds, 428
in *Uca,* 63, 64–67, 78, 92–94
Spawning in captivity, by *Sarda,* 272
Sphyraena, 438
Sphyraenidae, 297, 438
Spider crabs, 99
Spiny lobster, 25, 101, 463
Squirrelfish, 377, 381, 453
Staphylinidae, 1
Starling, 427
Stellar sea lion, 470–476, 487, 489
Stolephorus purpureus, 271
Stomach contents (*see also* Feeding), of *Katsuwonus,* 249
Stomatopods, 99, 103, 127, 140–155
Strombus gigas, 105
Submersibles, 251–253
Surgeonfishes, 278
Surgery, on pelagic fishes, 273
Swimming behavior
speed, in *Pomatomus,* 303, 305, 307–321, 324–325
in tunas, 250, 255, 267, 272–273
Synalpheus, 102

Talitridae, 1, 3, 42
Talitrus, 43
T. saltator, 4, 5, 6, 39
Talorchestia deshayesei, 4, 6, 39
T. longicornis, 4, 39
T. megalophthalma, 4, 39
Taxis
anemotaxis, in amphipods, 8
geotaxis
in decapods, 11
in isopods, 7–8, 43
in lycosid spiders, 8, 43
klinotaxis, in *Urosalpinx,* 163
menotaxis

Taxis (*cont.*)
in amphipods, 4
in lycosid spiders, 8
in *Uca,* 16, 33, 38
phototaxis
in isopods, 7
in lycosid spiders, 8
in *Urosalpinx,* 162
rheotaxis, in *Urosalpinx,* 162
telotaxis
in amphipods, 5
in *Uca,* 16, 29, 33, 43
Television, underwater, 251, 281, 440–441, 453
Tenebrionidae, 1, 9
Three-spined stickleback, 304, 305
Thunnus, 252
T. albacares, 250, 255, 258, 262, 264, 266, 272
T. obesus, 269, 271
Toadfish, 361–376, 379–383, 386–431, 439
Trichopsis vittatus, 377, 379
Turbidity, effect on feeding
in *Eupomacentrus,* 297
in *Phoca,* 489
Tursiops truncatus, 251, 436, 470
Tylidae, 1
Tylos latreilli, 7–8, 39
T. punctatus, 7–8, 39, 42, 43

Uca, 2, 11–38, 41–44, 60–61, 100
U. burgersi, 65
U. longisignalis, 65
U. maracoani, 62
U. minax, 63
U. princeps, 62
U. pugilator, 3, 10, 12, 14, 19, 25–38, 40, 62, 63, 65, 66, 68, 69, 71, 72, 74–79, 82, 85, 91–94, 102, 462
U. pugnax, 63, 65, 67
Uca (*cont.*)
U. rapax, 13, 19, 64, 65, 72, 73, 75–77, 93, 94, 101
U. speciosa, 65–67
U. spinicarpa, 65, 66
U. stylifera, 62
U. subcylindrica, 63
U. tangeri, 10, 12, 15, 18, 24, 25, 37, 40, 102
U. thayeri, 63, 93
U. virens, 65
Ultrasonic tags, to track tuna, 255
Urosalpinx cinerea, 157–243

Vocalization (*see* Communication, Sound detection, Sound production)
Vision (*see also* Eye)
acuity
in cetaceans, 469, 490
in land mammals, 487–488
in pinnipeds, 470–489
in tunas, 264, 271, 273
signals emitted, by *Ocypode,* 64
stimulus responses, of tunas, 258, 262, 273
use in feeding, in *Pomatomus,* 324

Wahoo, 260
Wolf spiders, 1, 3, 8–9, 11–12, 18, 2 25, 40, 42, 43
Weddell seal, 437
Whale shark, 259
Whitetip shark, 259
White-sided dolphin, 436, 469, 490
Wrasses, 278, 439

Yellowfin tuna, 250, 255, 258, 262, 264, 266, 272
Yellowtail snapper, 297, 439, 460

Zalophus californianus, 436, 470, 477–490